Abstract Machine

Humanities GIS

Charles B. Travis

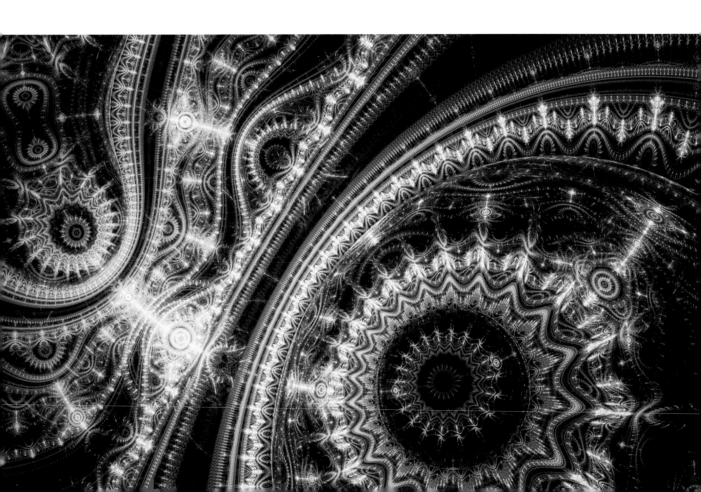

Esri Press, 380 New York Street, Redlands, California 92373-8100
Copyright © 2015 Esri
All rights reserved. First edition 2015

Printed in the United States of America

19 18 17 16 15   1 2 3 4 5 6 7 8 9 10

Library of Congress Cataloging-in-Publication Data
Travis, Charles, 1964-
    Abstract machine : humanities GIS / Charles B. Travis. -- First edition.
        pages cm
    Includes bibliographical references and index.
    ISBN 978-1-58948-368-2 (pbk. : alk. paper)—ISBN 978-1-58948-398-9 (electronic) 1. Geographic information systems. 2. Information storage
and retrieval systems—Humanities. 3. Humanities—Technological innovations. 4. Humanities—Methodology. 5. Space and time. 6. Geography and
literature. 7. Ireland—In literature. 8. Ireland—History. I. Title.
    G70.212.T74 2015
    001.30285—dc23   2014027420

Ask for Esri Press titles at your local bookstore or order by calling 800-447-9778, or shop online at esri.com/esripress. Outside the United States, contact your local Esri distributor or shop
online at eurospanbookstore.com/esri.

Esri Press titles are distributed to the trade by the following:

In North America:
Ingram Publisher Services
Toll-free telephone: 800-648-3104
Toll-free fax: 800-838-1149
E-mail: customerservice@ingrampublisherservices.com

In the United Kingdom, Europe, Middle East and Africa, Asia, and Australia:
Eurospan Group
3 Henrietta Street
London WC2E 8LU
United Kingdom
Telephone: 44(0) 1767 604972
Fax: 44(0) 1767 601640
E-mail: eurospan@turpin-distribution.com

Portions of chapters 2, 4, 6, and 8 are reproduced by kind permission of the following:

Springer Science + Business Media: *History and GIS: Epistemologies, Considerations, and Reflections* by Alexander von Lünen and Charles Travis, editors; chapter 12 "GIS and History:
Epistemologies, Reflections, and Considerations" by Charles Travis, p. 173–193, 2013.

Taylor & Francis Group, www.tandonline.com, "Transcending the cube: Translating GIScience time and space perspectives in a humanities GIS" by Charles Travis, *International Journal of
Geographical Information Science (IJGIS)*, vol. 28:5, p. 1149–1164, 2014; and from "From the ruins of time and space" by Charles Travis, *City*, vol. 17, issue no. 2, p. 209–233, 2013, article
DOI: 10.1080/13604813.2012.754191.

Edinburgh University Press, http://www.euppublishing.com/journal/IJHAC, "Abstract Machine—Geographical Information Systems (GIS) for literary and cultural studies: 'Mapping
Kavanagh'" by Charles Travis, *International Journal of Humanities and Arts Computing*, vol. 4, p. 17–37, available online October 2010, DOI 10.3366/ijhac.2011.0005, ISSN 1753–8548.

To my son, Senan James: like this book, you were a long time coming but well worth the wait!

Contents

Preface: Abstract Machine

Growing up, I was captivated by maps that accompanied pieces of fiction and the ways in which writers depicted actual and imagined places. In 2006, I was awarded a PhD for a thesis titled *Lifeworlds: Literary Geographies in 1930s Ireland*. My research examined how writers represented landscape, identity, and sense of place in Ireland during the early twentieth century. "Human geographers" as we conceive of them today did not exist during that period, so I employed writers as their proxies, informed by Ian G. Cook's observation that geographers and:

> the novelist have much in common. Both seek to portray the activities of people within the context of a specific milieu, infusing their descriptions of people and places with a sensitivity born of a rich and varied experience of life and society. Both seek to engender in their audience a deep awareness and empathy concerning others and their *lebenswelt*.[1]

My work was not a literary study per-se because it focused on phenomenology and the experience, perception, and representation of actual landscapes as captured through the prism of a writer's imagination. However, I became aware of the power of writing techniques and the strong relationship between literature, history, and place. I was also exposed to critical theory, particularly Mikhail Bakhtin's idea of the *chronotope*, a time-space motif that acts as a historical and geographical knot to tie a piece of literature's narrative strands together. At the time, I was teaching geographical information systems (GIS) in a graduate-level environmental science course and started to think about its convergences with literary approaches to place, in light of Gilles Deleuze and Félix Guattari's claim that "writing has nothing to do with signifying. It has to do with surveying, mapping, even realms that are yet to come."[2] Deleuze and Guattari also propose, "when one writes, the only question is which other machine the literary machine can be plugged into . . . in order to work."[3] In the context of the digital humanities, I thought that GIS might be a type of "machine" with which to do this. The creation and genealogy of GIS also seemed to conform to Deleuze and Guattari's definition of other human-technological interfaces comprising the forms and functions of an abstract machine:

> The double deterritorialization of the voice and the instrument is marked by a Wagner abstract machine, a Webern abstract machine, etc. In physics and mathematics, we may speak of a Riemann abstract machine and in algebra of a Galois abstract machine.[4]

Initially, the minimalist alliteration of point, polyline, and polygon-layer digital-mapping techniques employed in the abstract machine of a GIS did not captivate me. However, I soon began to consider whether Bakhtin's time-space nodes, and the narrative strands they tie together in a piece of literature, could be "plugged" into such an abstract machine and then correlated with latitude and longitude coordinates to map the relationships between actual and imaginary locations in a writer's depiction of place. Stuart Aitken and James Craine have observed that GIS acts not only as a technology of image making and communication but also as one of information transfer and knowledge production. They note that GIS comprises a chain of practices and processes through which users can gather geographical information and from it construct imaginative geographies.[5] The prospect of using GIS for digital mapping, spatial modeling, and storytelling in the humanities thus became even more intriguing to me.

And this is where I mix my metaphors.

For years, I played a Fender Stratocaster, an electric guitar that revolutionized the blues, jazz, country music, and rock and roll. Scoring and recording music is by its nature a mathematical and creative process, and innovations in sound engineering, which set musical inspiration to flight, could not have been made possible without discoveries in the fields of physics and electronics.

Musical artists such as Philip Glass and Peter Gabriel and bands such as Pink Floyd and U2 also intrigued me because they artistically engaged technology to compose their landscapes of sound. I was particularly fascinated by David Bowie's collaboration with Brian Eno, which produced the 1977 album *Heroes*. Bowie pointed out that the synthesizers used in the album had been designed by engineers, not musicians. The pair discarded their synthesizer manuals and simply played and experimented with the machines to create the sonic textures that shape their distinct soundscapes. This struck me as highly innovative in its simplicity and in turn influenced my own playful engagements and thoughts of employing GIS as a technology. Consequently, I follow a similarly idiosyncratic approach to historical, cultural, and literary GIS scholarship, in which I consult Esri tutorial manuals and then "critically play" with the software's digital suite of tools while keeping the tropes of the humanities firmly in mind.

One way to consider this method is to think about how music is charted and performed. Melodies and rhythms are schematically diagrammed on the staves according to mathematical principles. However, when translated by a musician, these representations create sonic vibrations in space, which cross the threshold from the quantitative to the qualitative, creating an entire liminal space of performance and reception.

In many ways, GIS has become my Fender Stratocaster.

I believe that GIS scholarship in the arts and humanities will proliferate by conceiving and developing its own unique languages, tools, and methodologies. We can therefore reconceptualize the operations of GIS as a creative suite in which to play and perform various digital-mapmaking and spatial-modeling techniques. In Esri's ArcGIS software, the ArcMap application constitutes a digital canvas upon which to plot abstracted layers of points, polylines, and polygons. The ArcCatalog application comprises a library platform from which the digital layers to create a map can be produced, borrowed, and returned. The animated mapping features of the ArcScene and ArcGlobe environments transform the possibilities for representing the dynamic relation between space and time in a cinematic manner, echoing Walter Benjamin's observation, in 1936, of the ability of film "to assure us of an immense and unexpected field of action . . . with the close-up, space expands; with slow motion, movement is extended."[6] Indeed, "time-space GIS movies" draw on the emotional power of moving images to tell visualized stories because cinematic techniques offer a kinetic visual experience more characteristic with the tropes of modernity to transform the possibilities for representing space in GIS.[7]

As a visualization tool most strongly associated with the discipline of geography (translated from the Greek for "earth writing"), we can conceptualize a humanities GIS model as an electronic stylus and digital cuneiform to inscribe spatial language and symbols. Such a model resurrects the practice of *geographia* from its ancient Greek and Roman roots, when the discipline constituted a literary genre more so than a branch of physical science.[8] When I teach GIS, I tell my students that to become proficient, they must practice their craft, just as if they were learning to play a piano, a saxophone, or a Fender Stratocaster. By approaching the "science" of GIS as an "art" form, humanities students can creatively play and experiment with its toolkit to translate its approaches for their own research and fields of study. This book offers a few creative examples of humanities GIS models, but the budding geospatial John Coltranes, Pablo Picassos, and Georgia O'Keefes are out there somewhere. I hope this book provides them with some inspiration.

Sources

1   I. G. Cook, "Consciousness and the Novel: Fact or Fiction in the Works of D. H. Lawrence," in *Humanistic Geography and Literature: Essays on the Experience of Place*, ed. D. C. Pocock (New Jersey: Barnes & Noble Books, 1981), 66.

2   G. Deleuze and F. Guattari, *A Thousand Plateaus*, trans. B. Massumi (London: Athlone, 1988), 4–5.

3   Ibid., 4.

4   Ibid., 141–42.

5   S. Aitken and J. Craine, "Affective Geographies and GIScience," *Qualitative GIS: A Mixed Methods Approach*, ed. M. Cope and S. Elwood (Thousand Oaks: Sage, 2009), 141.

6   W. Benjamin, *Illuminations* (London: Fontana Press, 1992), 229.

7   M.-P. Kwan, "Affecting Geospatial Technologies: Toward a Feminist Politics of Emotion," *The Professional Geographer*, 59, no. 1 (2007): 22–34; D. Cosgrove, "Maps, Mapping, Modernity: Art and Cartography in the Twentieth Century," *Imagi Mundi*, 57, no. 1 (2005): 35–54.

8   J. S. Romm, *The Edges of the Earth in Ancient Thought: Geography, Exploration, and Fiction* (Princeton: Princeton University Press, 1992), 3–4.

Acknowledgments

I would like to recognize the influence and encouragement of the following individuals: Professors Poul Holm, Anne Buttimer, Gunnar Olsson, Jane Ohlmeyer, David Nemeth, David Bodenhamer, Ian Gregory, Kevin Archer, and Stephen Reader; Drs. David Drew, Alexander von Lünen, Mary Gilmartin, Mark Hennessy, Stephen McCarron, Kieran Rankin, Krysia Rybaczuk, and Francis Ludlow; the poet Brendan Kenneally; and Gabriel García Márquez. I am grateful to Jo D'Arcy; Eadaoin Clarke and the Clarke family; Walter Price; my late father, Professor Charles Travis; my mother Kathleen Glavin Travis; and the Trinity Long Room Hub at Trinity College Dublin. I want to acknowledge Esri Press for shepherding the manuscript to production. I want to thank David Bowie, Brian Eno, Peter Gabriel, and the members of U2 for lessons in how creativity can bloom from the most unexpected places.

Part 1

GIS and the digital humanities

Chapter 1

Introduction

Figure 1.1 **The Lascaux Cave paintings (Hall of Bulls).** Photo courtesy of Jack Versloot (http://flickr.com/photos/80749232@N00).

From Lascaux to the Sea of Tranquility

The series of Paleolithic paintings of humans, animals, and cryptic signs that adorn the stone walls of the Lascaux cave complex in southwestern France (figure 1.1) bestow on humanity a 20,000-year-old tableau that we can appreciate, in an ontological sense, as one of the earliest humanities GIS models ever created. To literal minds, the images perhaps represent no more than the esoteric daubs of our ancestors huddled in fear around a primordial fire. To imaginative minds, however, the creative use of pigments, derived from local flora and fauna, illustrates a prehistoric knack for storytelling, artistry, and technological prowess. Indeed, these cave paintings, because they unambiguously plot the human-environmental interactions of hunter-gatherers, act as a primal GIS created to convey the "spatial stories" of a nomadic people still in thrall to the great myths and mysteries of the universe.

Fast-forward to the late twentieth century. We see *Homo sapiens* chart the solar system and navigate an Apollo rocket and lunar module to the Sea of Tranquility on the surface of the moon. Through its portal, we gaze back on an *earthrise* from the module's hi-tech cave. Now we cue the spool of history to the present to witness the digital revolution creating waves that ripple through the sciences, arts, and humanities. Amidst this great change, words such as *mapping* have emerged as important metaphors. In the arts and humanities, scholars navigate texts and explore the spatial and geographical dimensions of literary, cultural, and historical works.[1] Indeed, these spatial and cultural turns reveal that there are still many regions of *terrae incognitae* left to explore and map.

Drawing on tropes in the spatial and digital humanities, literary theory, and critical thought, this book illustrates how geographers can model and apply GIS techniques typically employed in the natural and social sciences to literary, cultural, and historical studies. This book takes the view that a humanities GIS model provides a discursive and artistic platform that we can use to visualize and spatialize stories and plot conventional empirical narratives. In addition, GIS can also be employed to perform ergodic and deformative interpretive mappings of literary, cultural, and historical works; create innovative, interactive digital texts; and foster insightful mapping experiences. This book targets students, researchers, and academics engaged in the digital humanities and anyone interested in how location, place, and space can illuminate their respective area of study.

What is a GIS?

For centuries, maps were sketched painstakingly by hand with materials such as ink, papyrus, sheepskin, parchment, and paper. The transition from traditional forms of mapmaking to interactive, digital mapping platforms, such as GIS, began in the late twentieth century. In our age, geocoded digital images proliferate, conjured on plasma screens by fingertip strokes on cybernetic keyboards that parse signals from earth-orbiting satellites. In the 1970s, computer mapping introduced the first digital maps and automated the drafting process from the sketching table to the computer screen. In the 1980s, electronic database systems linked to digital maps, which allowed the visual display of multidimensional data variables and provided the foundation for many GIS systems operating today.

A geographic information system, or GIS, provides a digital platform upon which multiple map layers (called *shapefiles* and *rasters*) electronically stack on top of each other to create composite images. Each shapefile layer and its attendant data table display unique variables (represented as points, polylines, and polygons). Layers can also be composed of a pixelated terrain or map images called rasters. The GIS operator digitally manipulates the order of the stacked layers and associated data tables, creating any number of connections between the spatialized variables to produce composite mappings, visual representations, and spatial models for analysis.

GIS software offers the potential to orchestrate, analyze, and visualize spatial stories as numerous as the shapefile and data variable combinations GIS operators can make. Collectively, as David Staley notes,

> Geographic information systems are one example of a suite of technologies—from data mining to immersive virtual reality displays to complex mathematical spaces—that have been collectively labelled "information visualizations." A visualization is any graphic that organizes data into spatial forms for purposes of display, analysis, interpretation, and communication.[2]

Traditionally, GIS technology has engaged Cartesianism and Euclidian geometry with positivist methodologies to create what the historian Michel de Certeau describes as "a formal ensemble of abstract places."[3] Similar to cartographic tools that translate the perceptible world onto a legible tableau, GIS constitutes an "abstract machine" designed to create conceptual spaces in which users can collate and then quantify, geocode, and visualize singular events and larger patterns to produce a qualitative "collage of moments."[4] The space-time backgrounds created by this combination of cybernetic systems and software languages produce a "qualculative" world in which calculation is defined as not necessarily being precise and super-computing technologies, qualitative choices, and ambiguity empower users to explore place and write space in different ways, both literally and metaphorically.[5] In this new world, GIS can be configured for use beyond positivistic endeavors and applied with innovation and imagination to the *terrae incognitae* of the humanities.

In this regard, a humanities GIS model provides a phenomenological tool that brackets events and patterns in both time and space. GIS can help users devise methodologies that are both quantitative (plotting geometric and numerical data relationships) and qualitative (juxtaposing attribute data relationships) to tackle important questions in literary, cultural, and historical studies. (See examples of the former in chapter 3 and the latter in chapters 4, 5, 6, and 7.) Whether a GIS is employed for quantitative or qualitative research, the selection of data involves a high degree of subjectivity—a trope with which arts and humanities scholars are conversant. Using GIS, these scholars can employ a spatial lens and apply many perspectives and analyses to any given subject by combining shapefile layers, data variables, and methodological approaches gathered from the sciences, the arts, and the humanities.

In the preface, GIS is defined as an "abstract machine." For the humanities, the significance of this abstract machine lies in its potential to contextualize the system's hardware and software nexus in a discipline that employs human and electronic cybernetic systems to advance our understanding of physical and social systems.[6] According to Donna Haraway, the cybernetic perspective of the late twentieth century emerged when innovators began to theorize human interactions with technology and fabricate the machine-organism hybrids called cyborgs.[7] This perception encouraged a mass proliferation of cybernetic assemblages that technologically disrupted Western ontologies and epistemologies. Far from being deterministic or dystopian, however, "cyborg imagery can suggest a way out of the maze of dualisms in which we have explained our bodies and our tools to ourselves . . . it means both building and destroying machines, identities, categories, relationships, space stories."[8] At its basic level, GIS constitutes a language of abstractions—a spatialized and cyber syntax articulated by cyborg authors who digitally reproduce perceptions of physical and social systems in codescapes of algorithmic, computerized commands.

This capacity suggests a cybernetic map—to parse Gilles Deleuze and Félix Guattari's aphorism, in which "there is no longer the tripartite division between a field of reality (the world) and a field of representation (the book)"—or the map—"and a field of subjectivity (the author)"—or the mapmaker.[9] The digital architecture of GIS software can be considered as an abstract machine that deterritorializes an individual's phenomenological sense of place by geometrically projecting idiosyncratic perceptions of the environment onto a coordinated grid system. The result is an abstract space that can be navigated, mapped, and studied. This is the dominant spatial perspective, epistemology, and methodology employed by most GIS practitioners. However, from a post-structural perspective, GIS can be conceived as a topographical hermeneutic system operating on an inter-textual platform that employs a spatial and cyber syntax to produce the postmodern notion of a digitally visualized mapping text. As discussed more fully in chapter 2, we can employ GIS to explore and survey "rhizomatic" spatial relationships and networks linking literary, historical, and cultural scales and networks on multidimensional levels (figure 1.2).

In Deleuze and Guattari's terminology, GIS constitutes an abstract machine that scholars use to pilot through space that does not represent something real so much as it constructs a reality that is yet to come.[10] This concept suggests not only an image-making technology but, more importantly, a technology that facilitates information transfer, knowledge production, and communication. GIS orchestrates a chain of practices and processes that gathers geographical information and constructs imaginative geographies.[11] The paths that a humanities-oriented GIS may follow, therefore, are many.

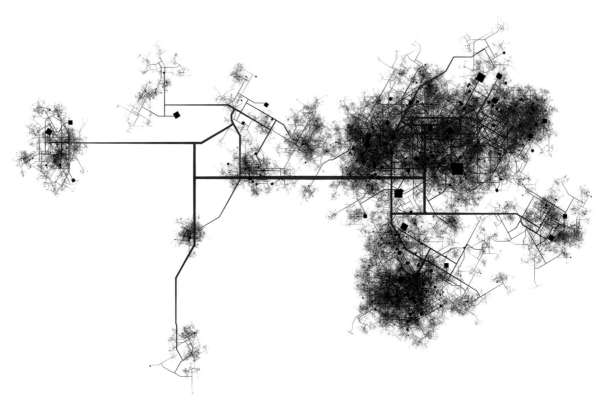

Figure 1.2   **Rhizome—the space of conjecture, as envisioned in a transport map blueprint of a city illustrating "rhizomatic" spatial relationships and networks.** Courtesy of Kentoh/Shutterstock.com.

GIS and the digital humanities

Scholars in the humanities have explored the relationship between geography and literature from cartographical and theoretical perspectives. Key examples range from Franco Morretti's (1998, 2005) schematic and Marxist geometrical approaches to literary studies to Barbara Piatti's ongoing project to map the fictional and actual locations of literary works.[12] Bertrand Westphal's "geocritical" approach explores the overlapping territories of physical geography, cognitive mapping, and literature by plotting the geometric and philosophical coordinates of real and fictional space through the conceptual lenses of *spatiotemporality*, *transgressivity*, and *referentiality*.[13] At the same time, a number of edited volumes have helped to further engage GIS in the humanities in both practical and theoretical ways and deepen the perceived connections between cartography, theory, and literature.[14]

Building on these works, I have created humanities GIS models that combine and perform ergodic and deformative readings of Irish literary, cultural, and historical texts.

Staley notes that in a typical written narrative, elements of a story operate in a linear pattern, with a beginning, middle, and an end. A spatial narrative can have a linear pattern (as in tour-style maps), but other forms of meaningful patterns can unfold in two or more dimensions. By departing from classic Aristotelian linear narrative, ergodic approaches in GIS—those that require work from an author-reader—can model complex spatial relationships between the author's construction of a text, the tabulation of archival data, and a viewer's choices.[15] This type of storytelling in GIS provides an interactive platform from which to synchronize layers of images, words, numbers, and vectors into simultaneous and multidimensional narratives.[16]

In contrast to ergodicity, *deformance* is a literary-criticism technique developed in the digital humanities as a key methodology for textual analysis and data mining.[17] The approach combines two words, *performance* and *deform*, to construct an interpretative concept premised on deliberately misreading a text—for example, reading a poem backward line by line.[18] In *Reading Machines* (2011), Stephen Ramsay notes that computers enable scholars to practice deformance quite easily—to take apart an epic poem, for example, by focusing only on its nouns or by calculating the frequency of collocations between character names in a novel.[19] Chapters 5, 6, and 7 of this book specifically test GIS-framed, deformative mapping models of the texts and biographies of three authors. Jerome McGann and Lisa Samuels contend that this interpretative technique applies *scientia* to *poiesis* to elucidate the relationship between two discourse forms. Furthermore, they argue, this method seeks to explain a unitary and unique phenomenon, rather than establish a set of general rules or laws.[20]

In addition to the application of ergodic and deformative techniques, this book situates humanities GIS in the fields of multimedia art, design, and culture. Here, according to Andrew Mactavish and Geoffrey Rockwell, humanities computing falls in league with the visual and performing arts by legitimizing technological practice and the creation of non-textual scholarly artifacts.[21] The use of GIS in this context illustrates Alan Liu's point: beyond acting in an instrumental role, the digital humanities broaden the very idea of instrumentalism, technological and otherwise.[22] Lev Manovich predicts that the systematic use of large-scale computational analysis and interactive visualization of cultural patterns (made possible with GIS) will grow into a major trend in cultural criticism and the culture industries in the coming decades. Manovich asks: "What will happen when humanists start using interactive visualizations as a standard tool in their work, the way many scientists do already?"[23]

Contents

This book presents a series of case studies related to the creation of humanities GIS models that blend tropes from literary, cultural, and historical studies. Reconceptualizing GIS by offering these types of ontological translations will hopefully foster epistemological marriages between qualitative and quantitative (or mixed-method) approaches; provide a means to visualize literary interactions with place and space, as well as critical theory; and creatively engage technological applications relevant to the digital humanities. The models used essentially implement selected GIS applications from my PhD dissertation on the "lifeworlds," or literary geographies, of 1930s Ireland and a GIS database-mapping project on seventeenth-century Irish land transfers as a postdoctoral fellow in the digital humanities.[24] Using GIS in this new way, I discovered that I enjoyed the process of mapping and gained insight from the experience—which brings to mind the aphorism that it is the journey, not the destination, that counts. I hope this book will inspire readers to undertake their own journeys in humanities GIS as well.

Part 1, "GIS and the digital humanities," continues with chapter 2, a brief history of Western geographical thought and post-structuralist theory in relation to the conceptualization of GIS approaches for the humanities. It ends with chapter 3, which presents a historical GIS case study of seventeenth-century Ireland that illustrates how three-dimensional (3D) geovisualization and database-mapping techniques helped me to analyze the redistribution of confiscated land following the 1641 Rebellion and Oliver Cromwell's 1649–50 conquest of Ireland.

Part 2, "Writers, texts, and mapping," focuses on GIS applications in literary and cultural studies. This section draws on the critical, aesthetic, and spatial thought of Mikhail Bakhtin, Walter Benjamin, Henri Lefebvre, Giambattista Vico, and the poet Dante. Chapter 4 draws on digital fieldwork to chronotopically plot the rural and urban landscapes experienced and perceived by the writer Patrick Kavanagh. This chapter chronicles how his relocation from the country to the city influenced the contrasting topophilic and topophobic depictions of his native Inniskeen Parish in his 1938 novel *The Green Fool* and 1942 epic poem *The Great Hunger*. Chapter 5 uses GIS to deformatively and ergodically reconstruct James Joyce's *Ulysses* and illustrate how the spatial influence of the medieval Italian poet Dante, in the words of poet Ezra Pound, inspired Joyce to launch a "new *Inferno* in full sail."[25] Joyce employed cartographical and artistic methods of the Cubists and Italian Futurists (among others) to plot his masterwork, so this chapter considers what may have been further accomplished had he access to current GIS tools.

Through the prism of GIS, chapter 6 explores the psychogeographies of Flann O'Brien's *At Swim-Two-Birds* (1939) and the historical cycles and poetics of Giambattista Vico and Mikhail Bakhtin to conduct Situationist International–inspired urban field surveys of modernist literature. Part 2 concludes with chapter 7, which discusses an open source–enabled GIS timeline created to perform an ergodic, digital bricolage that maps the writer Samuel Beckett's early life in Dublin, London, and France between 1916 and 1945.

Part 3, "Toward a humanities GIS," features chapter 8, which argues for rebooting GIS so that we may begin to engage the concepts and tools of the humanities. This book focuses largely on modeling GIS practices inspired by the humanities, which are currently pioneered by a coterie of scholars from across the academic spectrum and very much open to further exploration and development.

Sources

1   P. Booker and A. Thacker, *Geographies of Modernism: Literatures, Cultures, Spaces* (London and New York: Routledge, 2005), 1.

2   D. J. Staley, "Finding Narratives of Time and Space," in *Understanding Place: GIS and Mapping across the Curriculum*, eds. D. S. Sinton and J. J. Lund (Redlands: Esri Press, 2007), 36.

3   M. de Certeau, *The Practice of Everyday Life*, (Berkley: University of California Press, 1988), 121.

4   T. M. Barnes, J. Corrigan, and D. J. Bodenhamer, eds., *The Spatial Humanities: GIS and the Future of Humanities Scholarship* (Bloomington and Indianapolis: Indiana University Press, 2010), 174.

5   P. Merriman et al., "Space and Spatiality in Theory," *Dialogues in Human Geography,* 2, no. 1 (2012): 19; for a further discussion of "qualculativeness," please see N. Thrift, "Movement-Space: The Changing Domain of Thinking Resulting from the Development of New Kinds of Spatial Awareness," *Economy and Society, 33,* no. 4 (2004): 582–604.

6   J. E. Dobson, Reply to comments on "Automated Geography," in *The Professional Geographer,* 35 (1983): 351.

7   D. Haraway, *Simians, Cyborgs and Women: The Reinvention of Nature*, (London and New York: Routledge, 1991), 151.

8   Ibid., 181.

9   Deleuze and Guattari, *A Thousand Plateaus*, 23.

10   Ibid., 142.

11   Aitken and Craine, "Affective Geographies and GIScience," 141.

12   F. Moretti, *Atlas of the European Novel 1800–1900* (London: Verso, 1998); Moretti, *Graphs, Maps, Trees: Abstract Models for a Literary Theory* (London: Verso, 2005); B. Piatti, *Die Geographie der Literatur: Schauplätze, Handlungsräume, Raumphantasien* (Göttingen: Wallstein Verlag, 2008); (Ein Literarischer Atlas Europas [Literary Atlas of Europe][http://www.literaturatlas.eu/en/]).

13   B. Westphal, *Geocriticism: Real and Fictional Space*, trans. R. T. Tally, Jr. (New York: Palgrave MacMillan, 2011); R. T. Tally, Jr., *Geocritical Explorations: Space, Place, and Mapping in Literary and Cultural Studies* (New York: Palgrave MacMillan, 2011).

14   *Placing History: How Maps, Spatial Data, and GIS Are Changing Historical Scholarship*, eds. A. K. Knowles and A. Hiller (Redlands, CA: Esri Press, 2008); *The Spatial Humanities: GIS and the Future of Humanities Scholarship*, eds. D. J. Bodenhamer, J. Corrigan, and T. Harris (Bloomington: Indiana University Press, 2010); and *GeoHumanities: Art, History, Text at the Edge of Place,* ed. M. Dear et al. (New York: Routledge, 2011).

15   Staley, "Finding Narratives of Time and Space," 45.

16   Ibid.

17   M. Sample, "Notes towards a Deformed Humanities," *Sample Reality* (blog), May 2, 2012, http://www.samplereality.com/2012/05/02/notes-towards-a-deformed-humanities/.

18   Ibid.

19   Ibid.

20   J. McGann and L. Samuels, "Deformance and Interpretation," *New Literary History,* 30, no. 1, "Poetry & Poetics" (Winter, 1999): 25–56.

21   A. Mactavish and G. Rockwell, "Multimedia Education in the Arts and Humanities," *Mind Technologies: Humanities Computing and the Canadian Academic Community,* eds. R. Siemens and D. Moorman (Calgary: University of Calgary Press, 2006).

22   A. Liu, "Where Is Cultural Criticism in the Digital Humanities?" in *Debates in the Digital Humanities*, ed. M. Gold (Minneapolis: University of Minnesota Press, 2012), http://dhdebates.gc.cuny.edu/debates/text/20.

23   L. Manovich, "How to Follow Global Digital Cultures, or Cultural Analytics for Beginners," in *Deep Search: The Politics of Search beyond Google*, eds. F. Stalder and K. Becker (Piscataway, NJ: Transaction Publishers 2009).

24   C. Travis, *Lifeworlds: Literary Geographies in 1930s Ireland* (Dublin: Trinity College Dublin, 2006).

25   Ezra Pound, letter to Homer Pound dated April 20, 1921, in *Pound/Joyce: The Letters of Ezra Pound to James Joyce, with Pound's Essay on Joyce*, ed. F. Read, (New York: New Directions, 1967), 189.

Chapter 2

Toward the spatial turn

Figure 2.1 **The geographer's science and storyteller's art (Homer).** Courtesy of Vasileios Karafillidis /Shutterstock.com.

A brief history of Western geographical thought

In *Geographies of Modernism* (2005), Andrew Thacker and Peter Brooker note the recent, strong interest in the geographical and spatial dimensions that shape and are represented in literary and cultural texts.[1] Arguably, the roots of Western geographical thought and cartographical practice spring from the entwined branches of literary, artistic, and scientific pursuits; etymologically speaking, the word *geography* derives from the Greek word for "earth writing" (*geo* = "earth," *graphein* = "writing"). With imagination, humanities scholars can engage GIS as a prosthetic device to survey the *terrae incognitae* of historical, aesthetic, and cultural "textual spaces." A brief history of Western geographical thought and cartographical practice reveals few precedents for reconciling these spaces and their diverse artistic, narrative, and scientific genealogies in a humanities GIS model.

Herodotus (484–25 BC) and Strabo (63 BC–ca. 24 AD) sifted through a vast storehouse of travelers' tales, separating fact from fiction and retelling the stories they thought were credible enough to claim a reader's attention.[2] In doing so, they contributed to defining one branch of geography as a narrative and idiographic discipline anchored in ethics and politics. In contrast, Eratosthenes (276–195 BC) and Ptolemy (90–168 AD) employed mathematics to calculate the spherical nature of the earth and develop early map projections that helped to establish a branch of geography as a type of "geometry with names,"[3] foreshadowing modern attempts to establish nomothetic practices in cartography and geography. Initially, "the geographer's science and storyteller's art could not be fully detached from each other."[4] Soon enough, a clash developed between them:

> An academic controversy was waged over the reliability of geographical data in Homer's *Odyssey*. Strabo, who believed the *Odyssey* to be authentic and reliable, in a long and controversial passage leveled criticism against Eratosthenes for holding that Homer should be read as a poet and not as a scientific authority.[5]

Tensions and debates between the discipline's positivistic and poetic strands shaped the evolution of Western geographical thought and cartographical practice. A humanities GIS model may provide common ground.

Michel de Certeau (1925–86) observed that medieval and early modern mapmakers narrated spatial stories and histories by mapping religious pilgrimages and crusades, which they often experienced personally. They embellished their maps with the traces of footprints and alongside them illustrated the successive events that took place in the course of their journeys, such as meals, battles, mountains, and river crossings. In this regard, mapmakers acted as tour guides, visually interpreting their experiences and the plots of navigators with traditional geographical knowledge (such as Ptolemy's *Geography*) into representations conveying the reason and manner for which the maps were created (figure 2.2).

However, de Certeau notes, "between the fifteenth and seventeenth centuries, the map became more autonomous"[6] and eliminated such pictorial embellishments. Transformed by Euclidian and descriptive geometry, maps began to constitute an ensemble of "abstract spaces," in which a unitary sense of projection subsumed the mapmaker's experience and traditional

Figure 2.2 **The North Sea—medieval spatial stories (map fragment, 1572).** Courtesy of Sergey Mikhaylov/Shutterstock.com.

geographical knowledge within its theatrical frame. Now, maps collate "on the same plane heterogeneous places, some *received* from a tradition and others *produced* by observation."[7] Over time, by erasing the itinerary and idiosyncratic perspective of the human tour guides who created it, the map began to "colonize space," producing "a totalizing stage on which elements of diverse origin" were "brought together to form the tableau of a 'state' of geographical knowledge."[8]

In short, between the fifteenth and seventeenth centuries, abstract projections gradually effaced the idiosyncratic perspective and experience of the mapmaker.

During this period, John Pickles notes, two scopic regimes dominated the arts and sciences in Europe. One developed through the use of Cartesian perspectivalism, which allowed 3D spaces and linear perspective to be depicted together in two dimensions on a flat surface. This regime fixed visual representation within a coordinated grid.[9] In 1425, the Italian artist Filippo Brunelleschi performed a "feat of magic" in a cathedral piazza in Florence when he demonstrated the optical illusion created by the vanishing point in a painting. His innovative technique had "irreversible implications for the entire future of western art."[10] The other scopic regime, associated with the descriptive school in seventeenth-century Dutch art, sought to develop techniques for representing a phenomenological view of the world in two dimensions.[11] This fostered a "mapping impulse" in both Dutch painting and cartography concerned with representing the world in an intelligible and accessible manner for the public good.[12]

Perspectivalism offered mapmakers a homogeneous space regulated by a grid-like network of coordinates that created theatrical "scenographic" representations. According to Pickles, particular forms of parametric space, geometry, and scale anchored this modern cartographic gaze and subjected cartography to a controlling epistemological perspective that miniaturized and universalized the world's enormous complexities for discrete purposes.[13] In distinction to cartography, however, geography remained a wide field of knowledge until and during the Enlightenment, R. J. Mayhew asserts, rather than a closed and controlling discipline. Because geography and literature acted as more permeable categories in the eighteenth century, many geographers continued to pursue interests comparable to that which we now consider literary.[14] This flexibility set a precedent for scholars working to conceptualize a GIS model that creatively synthesized approaches in the humanities and sciences. However, as Brian Harley observes, Western cartographical practices since the seventeenth century have propagated a standardized scientific model of knowledge that produces a "correct" relational model of the terrain being mapped.[15] Consequently, cartography became a scientific, empirically based profession that embraced the practices of classification, quantification, and instrumentation.

Over the next two centuries, governments, states, and nations established institutions to conduct surveys and produce maps and topographical records of the territories and colonies they possessed and controlled. Statistical mapping grew into an important tool for government bureaucracy, and map use and interpretation practices widely disseminated to schools and other institutions for the sake of social regulation.[16] During the nineteenth century, geography lost some of its permeability with other fields as it became institutionalized as a discipline in Britain, Europe, and the United States. Along with cartography, geography began to serve the interests of nation and empire in utilitarian and ideological manners. The "Darwinian Revolution" and "Neo-Lamarckianism" (beliefs about the organic roots of genetic factors in the environment) influenced the development of "environmental determinism," a now discredited pseudo-scientific geographical perspective arguing that peoples' regional environments determined racial and cultural differences. Meanwhile, as Neil Smith observes, the conflation of geography, cartography, and the ideology of imperialism shaped British geographer Halford Mackinder's concepts of the "world-island" and the "geographical pivot of history" as well as US historian Frederick Jackson Turner's "end of the frontier" thesis on the settlement and closure of the American West. However, such perspectives also raised a few questions. If geographers had entirely mapped, enumerated, and described all the world's cultures, territories, and nations—"relegating geography to the realm of the fixed"—then what purpose and further utility could the discipline offer? Indeed, fears of "the end of geography" at the end of the nineteenth century, fostered by a chimerical belief in the "closure of absolute space provoked powerful ideological effects" in many political and academic circles.[17]

As Smith notes, the era spanning the 1880s through the early 1920s stands out as a time of political and economic turmoil and a transformation in culture and science. These four decades marked unprecedented creativity and the development of new concepts and ways of considering the world.[18] During this period, modernist literature, music, art, and architecture developed, and, just as profoundly, the apprehension and depiction of space and time became inseparably interlinked, notes Henri Lefebvre. The shock waves of this seismic cultural shift first reverberated through intellectual and artistic spheres, where the old "clock-work universe" *formulae* of space and time dissolved in the face of Einstein's mind-bending theory of relativity. In the works of Paul Cézanne and the school of analytical cubism, perceptible space and perspective disintegrated as the line of horizon disappeared from paintings.[19] Pablo Picasso's 1937 painting *Guernica* illustrates this shift in perception (figure 2.3).

In geography, Carl Ortwin Sauer (1889–1975) and the Berkeley School of Cultural Geography developed a new methodological lens based on a cultural concept of landscape that rejected the idea of environmental determinism. Sauer's morphological studies of regions and societies placed particular emphasis on the temporal dimension of a panoramic lens:

> We cannot form an idea of landscape except in terms of its time relations as well as it space relations. It is in continuous process of development, or of dissolution and replacement.[20]

The school emphasized a synchronic approach to researching and mapping historical, cultural, and physical landscapes. Although empirically oriented in his methodology, Sauer

Figure 2.3 **The ruins of space and time: Czechoslovakian postage stamp of Picasso's *Guernica*.** Stamp image courtesy vvoe/Shutterstock.com. Stamp shows *Guernica* painting by Pablo Picasso from Museo Nacional Centro de Arte Reina Sofia (ca. 1967).

recognized the significant role that subjective perception played in creating distinct "senses of place" rooted in the phenomenological symbiosis existing between particular regions and cultures.

Nevertheless, twentieth-century cartography remained anchored in Euclidean geometry, even as intellectual and artistic praxes moved toward Einsteinian concepts of time-space.[21] These parallel tracks, which resembled the approaches of Cartesian perspectivalism and the seventeenth-century Dutch mapping impulse, manifested as two distinct schools of geographic practice. The former gave birth to the quantitative revolution of the 1950s and 1960s, with its emphasis on spatial modeling and computing, while the latter shaped geography's second cultural turn during the 1970s and 1980s, during which humanistic and postmodern scholars applied the metaphor of *text* to the acts of reading landscapes, conducting fieldwork, and framing social life.[22]

Post-structuralist perspectives

During this period, the word *mapping* emerged as a significant metaphor in the arts and humanities as scholars began to show strong interests in the roles of place, space, and the implicit geographical dimensions of literary and cultural texts.[23]

Emphasizing the spatial and postmodern trends emerging in the 1980s, Roland Barthes defined the word *text* as "a multi-dimensional space in which a variety of writings, none of them original, blend and clash. The text is a tissue of quotations drawn from the innumerable centres of culture."[24] The cross-pollination of these two methodological metaphors across disciplinary boundaries informed cartographic historian J. B. Harley's seminal observation:

> "Text" is certainly a better metaphor for maps than the mirror of nature. Maps are a cultural text. By accepting their textuality we are able to embrace a number of different interpretative possibilities. Instead of just the transparency of clarity we can discover the pregnancy of the opaque.[25]

Despite the new semiotic approach to studying landscape, by the end of the twentieth century, developments in computer science allowed GIS to become the indispensable tool for geographical research and analysis in government, business, and academia.[26]

This development in the arts and humanities coincided with the unprecedented phenomena of digital globalization facilitated by visual broadcast media and the World Wide Web. Marshall MacLuhan's observation in 1964 that, after more than a "century of electric technology, we have extended our central nervous system itself in a global embrace, abolishing both space and time as far as our planet is concerned," was radically prescient.[27] The ubiquitous use of personal computers, tablets, and smartphones; consumption of 24-hour mass-media outlets; and proliferation of social media profoundly shaped twenty-first-century geographic perceptions and practices. GIS, GPS, computer cartography, and online open-source geospatial software are framing the earth as a geocoded world that is continuously being coded, decoded, and recoded as new cybernetic language systems and platforms emerge and evolve.[28] Manuel Castells' argument that the geography of the new history will be constructed out of the interface between places and flows seems remarkably apt.[29] Contemporary human geographical practices are engaging space as a dynamic "lifeworld" and a "quasi-material construct" produced by social interaction. Rather than a passive container, today space is increasingly considered as an active agent, infused with human behavior and perception, which is constantly shaping, producing, and reproducing places socially, politically, and economically.[30]

Recently, Nigel Thrift has promoted nonrepresentational theory as an experimental perspective concerned with *the geography of what happens*. This approach pulls the vibrant energy of the performing arts into the social sciences by crawling out on the edge of a conceptual cliff. Thrift proposes that, because of the intervention of software, the human body has become a tool-being in symbiosis with a new electronic time-space that shapes our perceptions and experiences of the world, echoing Donna Haraway's cyborg theory.[31] Therefore, the current geographical concern with human performativity and dynamic social space, and their relationships to "automated" mapping functions, cyber linguistics, and Web 2.0 (social media)/3.0 (semantic, geosocial, and 3D visualization) platforms, can facilitate deeper and more experimental forms of GIS engagement with research and scholarship in the arts and humanities.

Deep mapping

As a result of the spatial turn in the humanities, the word *mapping* became a discursive metaphor that scholars employed to discuss the spaces and places that shaped literary and historical texts and were represented in them. By establishing humanities GIS models, scholars can now explore, survey, chart, map, and navigate textual journeys in a literal sense. Traditionally, under the umbrella of spatial science, users employed GIS to map distributions and patterns but engaged little in mapping the lyrical, subjective, temporal, and esoteric notions of place and human experience. The predominant positivistic perspective encounters a greater challenge in performing qualitative analysis and visualizing critical theoretical relationships in GIS, as well as captures the dynamic nature of evolving networks, the cascade of historical events, and the myriad social flows that interlink people and places.[32]

By adopting postmodern approaches, however, we can use GIS technology to create unique opportunities to construct alternate constructions of history and culture that embrace multiplicity, simultaneity, complexity, and subjectivity.[33] By placing historical and cultural exegesis more explicitly in space and time, GIS can identify patterns, facilitate comparisons, enhance perspectives, and visualize data in any number of ways. Currently, humanities-influenced methodologies are taking GIS technology beyond the limits imposed by positivism. Examples include "deep-mapping" techniques and approaches that engage oral cultures and storyscapes, detect semiotic traces on landscapes, and provide experiential interfaces with virtual data and immersive geospatial environments.

The writer William Least Heat Moon (William Trogdon) first conceptualized deep mapping as a vertical form of travel writing in his book *PrairyErth* (1991). This approach both records and represents the grain and patina of place through the interpenetrations and juxtapositions of the past and the present, the political and the poetic, and the discursive and the sensual.[34] David Bodenhamer asserts that using GIS deep-mapping methodologies to fuse qualitative and quantitative data acknowledges the reality of multi-scalar and dynamic space-time.[35] GIS techniques can assemble and visualize layers with different degrees of transparency to integrate oral testimony, anthology, memoir, biography, and natural history.[36] Similarly, Mark Palmer, in his work on Kiowa oral culture and storyscapes, coined the neologism *indigital GIS* to describe networks of indigenous, scientific, and technological knowledge systems that engage science, symbols, and stories to create fragmentary and contradictory geo-narratives full of uncertainties.[37] Researchers phonetically coded Kiowa and Latin alphabets into geodatabases to create maps that trace everyday paths by combining remote-sensing images and data from the Internet. This shape-shifting through GIS allows users to dramatize Kiowa oral culture and storyscapes that present long-elapsed events that unfold as if before one's eyes and summarily strips away a type of historical theater of the past.[38] By integrating Kiowa language, perceptions of terrain, seasons, the solar cycles, and the Milky Way in a dynamic and holistic fashion as geo-narrative art, Palmer's GIS storyscape techniques—with its many kinds of fusion, interbreeding, and boundary crossing—illustrate the direction of digital mapping.[39]

In addition, Wolfgang Moschek and Alexander von Lünen have engaged GIS as a means to semiotically track and interpret the ruins of limes, ancient Roman border fortifications located in Britain, as *signs* of a cultural mentality inscribed on the landscape to delineate perceived "civilized" and "barbaric" spaces.[40] Von Lünen argues that while positivists interpret such sources as open windows into the past, postmodernists perceive them as fences obstructing vision.[41] Modeling semiotic data in GIS resembles the early stages of tracking, through which a postmodern detective (fictional or actual) identifies and interprets clues to understand how their sources speak.[42] Once clues are assembled and parsed, GIS is engaged—not as a cartographical tool but rather as the intuitive scratchpad of a bricoleur—to encode scanned archival maps with clues that semiotically transform the *signified* source into an active and present historical *signifier*. Semiotic GIS techniques allow the practitioner to *arrange* and *elicit* signifier-signified meanings and intentions in human traces (records, documents, artifacts), rather than simply *analyze* established historical narratives.[43]

Last, Trevor Harris has created a visualization-gaming platform called *The Cave*, which immerses users in GIS-rendered landscapes, such as nineteenth-century Morgantown, West Virginia, to facilitate a phenomenological experience of a different environment, period, and place. Similar to the holodeck from *Star Trek*, this experiential form of GIS projects 3D models of townscapes and terrains—sourced from cartographical, archival, scientific, census, and literary data—on the walls, floor, and ceiling of an enclosed space. Individual users can then navigate the virtual environment to explore the fully rendered visualization from their own perspectives. To foster a user's sensual embodiment of the streets of Morgantown two centuries ago, Harris integrates sound effects, such as a beating heart that quickens as one navigates near townscape locations identified by nineteenth-century writers and historians as places of perceived fear or danger.[44] This type of immersive, experiential GIS environment also enables individuals and groups to create their own forms of visualization by interacting experimentally with virtualized spatial data.[45]

The preceding examples of humanities-based applications illustrate that GIS is not a fixed or singular identity but instead a technology that, with creativity and innovation, we can reconceptualize and retool to conduct more qualitative, lyrical, artistic, esoteric, and phenomenological forms of research.[46] These examples also show how the strong influence of humanities disciplines on GIS innovation can provide a new and ontologically different reality to geography itself.[47] Furthermore, these GIS models considered in the context of Deleuze and Guattari's thought can provide a space of conjecture in which to reconcile the literary, artistic, and scientific roots of geography as well as provide a way to reimagine the potential of geospatial technology applications and research for humanities scholarship.

GIS and the space of conjecture

According to Deleuze and Guattari, both mapping and writing possess the power to anticipate and reimagine configurations of space, time, language, and culture, which have either been submerged by Cartesian space or yet to be perceived and represented. In their books, "one has the sense that there is only geography, nothing but geography: maps, planes, surfaces, strata, spaces, territories, transversals, etc."[48] Their conception of striated and smooth space links, respectively, to *arborescent* and *rhizomatic* forms of epistemology. Deleuze and Guattari use the first term to describe hierarchical, finite, and closed systems of thought and representation and invoke the rhizome as a curling, anarchic, subterranean plant root system to illustrate the interconnectivities that link society, writing, technology, and the human mind. Subsequently, a few creatively minded geographers have used this metaphor to manage the messiness of interactions and interconnections between human and physical systems.[49]

Pickles and Harley observed that cartography originally developed as a particularly controlling gaze, tied to certain forms of parametric space, geometry, and scale, which by the nineteenth century had developed into an empirical "scientific" practice anchored firmly by positivistic perspectives. However, Harley notes that the "steps in making a map—selection, omission, simplification, classification, the creation of hierarchies, and 'symbolization'—are inherently rhetorical."[50] With the advent of GIS technology, the infusion of humanities practices and discourses of postmodernism into the syntax of cartography has significantly changed the discipline to allow new concepts to develop. In this respect, GIS appears to function as a type of automated rhetorical tool. William Cartwright notes:

> Clicking icons, rather than remembering long, alphanumeric strings revolutionized the way in which users interacted with a package. To properly understand each of the elements in a geographical information package, a number of metaphors may have to be used if the complex nature of the real world is to be presented in simplified, understandable ways.[51]

Spatial analysis in GIS is generally qualitative, visual, and intuitive, despite its technology being insistently pigeonholed as a tool for solely quantitative applications.[52] In fact, a good portion of GIS attribute data is qualitative in nature—including names (such as owners of land parcels, businesses, and street addresses) and types or labels (such as roads, settlements, and soils). In most cases, this factor makes such types of attribute data unsuitable for quantitative analysis, so they are usually queried and logically manipulated by employing the SQL (structured query language) feature of GIS—a parsing tool closer to the study of philology than it is to physics. The performance of complex attribute queries in GIS requires more than just statistical or mathematical aptitude; it demands logical thinking and spatial imagination—skills the humanities can hone.[53]

However, quantitative skills are still important to the practice of GIS; its mastery relies on both literacy and numeracy. To fully harness a humanities GIS model to our research purposes, we must create new vocabularies of space to serve them. In *A Thousand Plateaus*, for example, Deleuze and Guattari proposed such a vocabulary and coined new terms and phrases, such as *assemblage*, *deterritorialization*, *lines of flight*, *nomadology*, and *rhizome/rhizomatics*, to describe spatial relationships and the ways we conceive people and other objects moving in space.[54] Observed in the context of the twenty-first-century digital revolution, an integrated, multidimensional GIS application compares to a standard cartographic map "as the internet [does] to a letter."[55] Online and desktop GIS provide unprecedented rhizomatic networking potential by employing the hyper-connectivity of the web to survey, chart, and navigate new and emerging configurations of space and time. As Umberto Eco observes, "the rhizome is so constructed that every path can be connected with every other one. It has no center, no periphery, no exit, because it is potentially infinite. The space of conjecture is a rhizome space."[56] Such a space can provide a way to consider and imagine how, in a humanities GIS model, the ancient literary, artistic, and scientific branches of Western geography, in tandem with the "three key referencing systems—space, time and language—might be engineered"—as John Corrigan phrases it—"in such a way that changes in one ripple into the others."[57]

Sources

1   P. Booker and A. Thacker, *Geographies of Modernism: Literatures, Cultures, Spaces* (London and New York: Routledge, 2005).

2   Romm, *Edges of the Earth*, 3–4.

3   G. Olsson, *Abysmal: A Critique of Cartographic Reason* (Chicago: The University of Chicago Press, 2007), 32.

4   Ibid.

5   J. K. Wright, *Human Nature in Geography* (Cambridge: Harvard Press, 1996), 11.

6   Certeau, *Practice of Everyday Life*, 121.

7   Ibid.

8   Ibid.

9   J. Pickles, *A History of Spaces: Cartographic Reason, Mapping and the Geo-Coded World* (London and New York: Routledge, 2006), 80.

10   Ibid, 84–85.

11   Ibid.

12   Ibid.

13   Ibid., 80.

14   R. J. Mayhew, *Geography and Literature in Historical Context: Samuel Johnson and Eighteenth-Century English Conceptions of Geography* (Oxford: School of Geography, 1997), 7, 43.

15   J. B. Harley, "Deconstructing the Map" *Cartographica,* 26, no. 2 (1989): 4.

16   Cosgrove, *"Maps, Mapping, Modernity,"* 37.

17   N. Smith, *American Empire: Roosevelt's Geographer and the Prelude to Globalization* (Berkeley: University of California Press, 2002), 14.

18   Ibid., 13.

19   H. Lefebvre, *Critique of Everyday Life*, vol. 3, from *Modernity to Modernism (towards a Metaphilosophy of Daily Life)*, trans. G. Eliot (London: Verso, 2005), 46.

20   C. O. Sauer, *The Morphology of Landscape* (University of California Publications in Geography, 1925), 36.

21   H. Lefebvre, *Critique of Everyday Life*, vol. 1, *Introduction*, trans. J. Moore (London: Verso, 1992), 46.

22   Pickles, *A History of Spaces*, 54.

23   Thacker and Brooker, *Geographies of Modernism*, 1.

24   R. Barthes, *Image: Music: Text*, trans. S. Heath (Glasgow: Fontana, 1982), 146.

25   Harley, "Deconstructing the Map," 7–8.

26   J. E. Dobson, "The Geographical Revolution: A Retrospective on the Age of Automated Geography," *The Professional Geographer,* 45 (1982): 431.

27   M. McLuhan, *Understanding Media: The Extensions of Man* (London and New York: Routledge, 1987), 3–4.

28   Pickles, *A History of Spaces*, 5.

29   M. Castells, "Grassrooting the Space of Flows," *Urban Geography,* 20, no. 4 (1999): 294–302.

30   Historically, ideas about the nature of space have been considered by Gottfried Wilhelm Leibniz (1646–1716), David Émile Durkheim (1858–1917), Martin Heidegger (1889–1976), Maurice Merleau-Ponty (1908–61), and Henri Lefebvre (1901–91). Geographers such as Gunnar Olsson, Anne Buttimer, Yi-Fu Tuan, Nigel Thrift, David Harvey, Doreen Massey, and Edward Soja have drawn from their works to inform their geographical methodologies.

31   N. Thrift, *Non-Representational Theory: Space/Politics/Affect* (London and New York: Routledge, 2008), 2, 10, 12, 18, 89.

32   M. Pavloskaya, "Theorizing with GIS: A Tool for Critical Geographies?" *Environment and Planning A,* 38 (2006): 2015.

33   D. J. Bodenhamer, "Creating a Landscape of Memory: The Potential of a Humanities GIS," *International Journal of Humanities and Arts Computing,* 1, no. 2 (2007): 102, 107.

34   M. Pearson and M. Shanks, *Theatre/Archaeology* (London and New York: Routledge, 2001).

35   D. J. Bodenhamer, "Beyond GIS: Geo-Spatial Technologies and the Future of History," in *History and GIS: Epistemologies, Considerations and Reflections*, eds. A. von Lünen and C. Travis (New York: Springer, 2012), 12.

36   Pearson and Shanks, *Theatre/Archaeology* (London and New York: Routledge, 2001), 65.

37   M. Palmer, "(In)Digitizing Cáuigú Historical Geographies: Technoscience as Postcolonial Discourse," in *History and GIS: Epistemologies, Considerations and Reflections,* eds. A. von Lünen and C. Travis (New York: Springer, 2012), 43.

38   Palmer, "(In)Digitizing Cáuigú Historical Geographies," 55.

39   Dennis Wood, *Rethinking the Power of Maps* (New York: Guilford, 2010), 111, quoted in Mark Palmer, "Theorizing Indigital Geographic Information Networks," *Cartographica: The International Journal for Geographic Information and Geovisualization,* 47, no. 2 (2012): 80–91.

40   A. von Lünen, "Tracking in a New Territory: Re-imagining GIS for History," in *History and GIS: Epistemologies, Considerations and Reflections*, eds. A. von Lünen and C. Travis (New York: Springer, 2012), 211–39.

41   Ibid.

42   A. von Lünen, "History and GIS: Epistemologies, Considerations and Reflections." Panel session, "Historians in Space," Graduate Conference in European History, Budapest, Hungary, April 25, 2013.

43   Lünen, "Tracking in a New Territory," 211–39.

44 T. M. Harris, L. J. Rouse, and S. Bergeron. "Humanities GIS: Adding Place, Spatial Storytelling and Immersive Visualization into the Humanities," *GeoHumanities: Art, History, Text at the Edge of Place*, eds. M. Dear et al. (London and New York: Routledge, 2011), 226–40.

45 T. M. Harris and P. Hodza, "Geocollaborative Soil Boundary Mapping in an Experiential GIS Environment," *Cartography and Geographic Information Science,* 38, no. 1 (2011): 20–35.

46 Pavloskaya, "Theorizing with GIS," 2014.

47 Aitken and Craine, "Affective Geographies and GIScience," 144.

48 Notably, *Anti-Oedipus, Kafka, A Thousand Plateaus,* and *What is Philosophy?,* M. A. Doel and D. B. Clarke, "Giles Deleuze," in *Key Thinkers on Space and Place*, eds. P. Hubbard, R. Kitchin, and G. Valentine (London: Sage, 2006), 104.

49 N. Chrisman, "Full Circle: More than Just Social Implications of GIS," *Cartographica,* 40, no. 4 (2005): 28.

50 Harley, "Deconstructing the Map," 11.

51 W. Cartwright, "Applying the Theatre Metaphor to Integrated Media for Depicting Geography," *The Cartographic Journal*, 46, no. 1 (2001), 28.

52 Pavloskaya, "Theorizing with GIS," 2011, 2013.

53 Ibid.

54 N. Gough, "Geophilosophy and Methodology: Science Education Research in a Rhizomatic Space" (unpublished chapter, prepared for UNESCO-SAARMSTE book project on methodologies for researching mathematics, science, and technological education in societies in transition, 2005), accessed February 1, 2011, http://www.bath.ac.uk/cree/resources/noelg_SAARMSTE_ch.pdf.

55 Ibid.

56 U. Eco, *Postscript to the Name of the Rose*, trans. W. Weaver (New York: Harcourt Brace Jovanovich, 1984), 57.

57 J. Corrigan, "Qualitative GIS and Emergent Semantics," in *The Spatial Humanities*, 85.

Chapter 3

Writing time and space with GIS: The conquest and mapping of seventeenth-century Ireland

Period, place, and GIS

In *The Landscape of Time* (2002), John Lewis Gaddis links the ancient practice of mapmaking with the three-part conception of time (past, present, and future) that many historians apply to their work. Both practices manage infinitely complex subjects by imposing abstract grids over them—in forms such as hours and days or longitude and latitude on landscapes or timescapes. Observing this, he asks, "What if we were to think of history as a kind of mapping?"[1] If the past is a landscape and history the way we represent it, and pattern recognition constitutes the primary form of human perception, Gaddis reasons, then history, from the epic to the simple narrative, seeks to discern patterns in much the same way that we would a landscape. This landscape metaphor accommodates varying degrees of complexity, not only as a reflection of scale but also for the information available at any given time concerning a particular landscape, geographical or historical.[2]

Incorporating Gaddis' metaphor into a humanities GIS model can reconcile the discursive and visual tropes Herotodus and Strabo established 2,000 years ago to tell stories about times, places, and events. This relationship between historical and geographical practices has not completely disintegrated through the ages. John Smith, in his *Generall Historie of Virginia* (1624), declared, "As geography without history seemeth a carcass without motion, so history without geography wandereth as a vagrant without habitation."[3] Situating the two disciplines as cornerstones of his conception of the modern university, Immanuel Kant (1724–1804) observed that "geography and history fill up the entire circumference of our perceptions; geography that of space, history that of time."[4] The publications of Paul Vidal de la Blache's *Tableau de la Géographie de la France* (1903) and Carl Sauer's *The Morphology of Landscape* (1926) signaled the twentieth-century reintegration of historical and geographical techniques into synthesized methodological frameworks.

Nowadays, human geographers and GIScientists recognize that time and space imply and integrate each other and that this creates a compelling disciplinary need to look to history to explore the irreversibly emergent nature of space-time complexes as they evolve.[5] Meanwhile, historians have been considering the emplacement of time, recognizing that location, place, and environment play significant roles in shaping human behavior, relations, and organization. They point to Thucydides' plague, Montesquieu's climate, and Turner's frontier as illustrations of nature's profound influence on history.[6] The historian P. J. Ethington observes that all human action presumes a location in both time and space. However, "the past cannot exist in time: only in space. Histories representing the past represent the place (topoi) of human action."[7] Considering this, scholars have reconceived historical interpretation as "the act of reading places, or *topoi*."[8] Knowledge of the past, Ethington argues, functions cartographically, literally speaking. As a result, he claims: "the incalculable volume of historical writing on all subjects should be thought of as a map because the past can only be known by placing it, and the way of knowing places is to map them."[9] Consequently, history has become the primary field in the humanities to focus its tropes through the lens of a GIS. Because historical scholars deal with complex processes in dynamic, nonlinear systems, argues historian Jack Owens, GIS provides the ability to organize a large number of variables and identify those most crucial to the stability and transformation of such systems.[10] In turn, the historical geographer Anne Kelly Knowles contends that GIS mapping methodologies as well as data and visualization techniques compel writers to think graphically and spatial thinkers to accept the subtle nature of historical texts. She observes,

> Much of the historical GIS being done today involves visualizing places themselves—the digital reconstruction of past landscapes. Some geovisualization strives for verisimilitude for its own sake, inspired by the beguiling realism of computer animation in cinema and gaming. But digital reconstruction of historical landscapes can also serve scholarly purposes.[11]

Arguably, the strongest evident rapprochement between geographical and historical practices now occurs in GIS scholarship. In one sense, GIS constitutes a visualized codescape of computer language and commands that provide tools to historians wishing to employ the epistemologies of *looking*, *querying*, and *questioning* associated with emerging geovisualization techniques.[12]

In this chapter, GIS adopts these approaches to create dynamic visual narratives of the seventeenth-century conquest of Ireland and a spatialized demographic study of the land redistribution that took place in its aftermath. This chapter is by no means a definitive work on this contentious period in Irish history; rather it demonstrates the possibility of integrating methodologies and techniques from GIS and geography to conduct innovative empirical research and visualize the spatial stories that give shape to not only the epic, but the episodic tableaux composing the tapestries of history.

Geovisualizing Irish history

In August 1649, Oliver Cromwell (1599–1658) led the New Model Army into Ireland. He sought to pacify the island and conquer the Catholic Confederacy established in Kilkenny following the Irish Rebellion of 1641. Originally, the rebellion erupted in defiance of New English and Scottish Protestant settlements and plantations. Over the next decade, however, vicious ethno-religious conflict pitted the Gaelic Irish and the largely Catholic, longtime-settled Old English population against the newly arrived English and Scottish Protestants. Consequently, Ireland roiled with a level of terror similar in magnitude to the Thirty Years' War in Germany (1618–48).

The roots of rebellion trace back to the 1550s in Counties Laois and Offaly when English policy in Ireland began to dictate the institutionalized confiscation of land. When this policy unleashed sectarian warfare across the island, the English Crown replaced rebellious Catholic Irish landowners with Protestants loyal to the religion and politics of the Crown. Although landlords were replaced, tenants were not. This created a combustible ethno-religious demographic mix. Under surrender and regrant agreements from the 1540s, Gaelic Chieftains renounced the Irish titles to their land in exchange for English titles under the protection of English Crown law. However, this was followed in 1586 and 1606 by the creation of two overtly colonial projects, the Munster and Ulster plantations. Distributing lands to settlers and natives loyal to the Crown, Ireland's governors pursued a radical colonial project that sought to build a replica of England in religion, language, and society.[13] The 1641 rebellion erupted as a backlash to this policy.

In response, Cromwell and his New Model Army, inflamed by the Puritan revolution in England, sailed to Ireland believing that disloyal Catholics should pay in blood and land for their massacre of Protestant settlers. His scorched-earth policy during the Siege of Drogheda in September 1649 signaled both Cromwell's intentions and the ferocity of his crusade. In nine months, the New Model Army brought Leinster, Munster, and Ulster into a single jurisdiction; by the end of 1650, the army had secured these provinces firmly under English Commonwealth control. East, north, and south lay in English hands; this left the Connacht region in the center-west for Cromwell's son-in-law to subdue. As Catholic landowners fled the slaughter, he placed them in "reservations" in Connacht where they lived in exile. To make matters worse for the old Catholic confederates, Cromwell granted veterans of his army the exclusive rights to a 3-mile wide strip of land that ran along Connacht's Atlantic coast from Limerick City to North Mayo. This effectively landlocked Connacht's Catholic landowners, barring them free access to ports and the fruits of the maritime economy, including fishing. By 1652, the subjugation of the Irish island was complete, and in 1653 Cromwell assumed the title "Lord Protector of the Commonwealth of England, Scotland, and Ireland."

The terrain of the island certainly influenced the course and outcome of these conflicts, given that its landscapes and the style of fortifications encouraged siege warfare. Countless 2D maps have charted the rebellion and Cromwell's campaign. Today, however, GIS applications can contextualize these events in 3D models of spatio-historical patterns overlaid on Ireland's topography. These make it possible to reveal the dynamic nature of these campaigns and help us more deeply explore the influence of location and environment on the course of the 1641 rebellion and the violent conquest of Cromwell's 1649–50 campaign that followed.

Rebellion and conquest in 3D

Esri's ArcScene application helped to create a dynamic visualization of the 1641 rebellion using a digitized copy of *A Map of Ye Kingdome of Ireland*. The map, published around 1655, most likely accompanied an English propaganda pamphlet promoting investment opportunities in Ireland. The map legend informs the reader that the map includes "perticular notes distinguishing the townes revolted, taken or burnt since the late rebellion."[14]

The visualization of the rebellion's events developed in the following manner:

1.   I imported a digitized copy of *A Map of Ye Kingdome of Ireland* into ArcMap as a raster image and georectified it (figure 3.1).

2.   I then created eight polygon shapefile layers in ArcCatalog and edited them in ArcMap to visualize the eight stages of the 1641 Rebellion as it spread across Ireland.

3.   After importing the data into ArcScene, I extruded the polygon layers in descending height over the rasterized map image. The highest layer covered the area in Ulster where the rebellion began, and the lowest layer designated the western and southern coastal regions where it ended (figure 3.2).

In the case of Cromwell's 1649–50 campaign, stills captured from ArcScene animated model visualizations, featuring a triangulated irregular network (TIN) of Ireland's topography, illustrate the movements of his New Model Army across the island's terrain.[15] The TIN, composed of nodes and lines with 3D coordinates (x, y, and z), was processed using a GIS function to create the digital data structure of the island's elevation, which was then imported into ArcScene.

The rebellion involved Ulster and the West of Ireland, however, the visualization still in figure 3.3 focuses on the trajectories of Cromwell's army as his forces expanded like a spider web across the east and south of Ireland. He marched from Dublin to Drogheda in September 1649, then double-backed to advance down the island's east coast to lay siege to Wexford and capture New Ross before extending his deadly tendrils of terror into Waterford.

The ArcScene platform provides an interactive 3D environment in which users can spatially rotate and tilt to dynamically probe, study, and illustrate the chronology of historical events across digitized terrains. The placement of colored time-space boxes in the ArcScene TIN model allows Ireland's topography to appear in 3D, bas-relief form, as the still in figure 3.4 illustrates. This is a useful function because the paths of Cromwell's armies were sourced from static 2D

Figure 3.1 *A Map of Ye Kingdome of Ireland* **(1655).** From *A Map of Ye Kingdome of Ireland* dated to 1655 in the British Library. Author unknown.

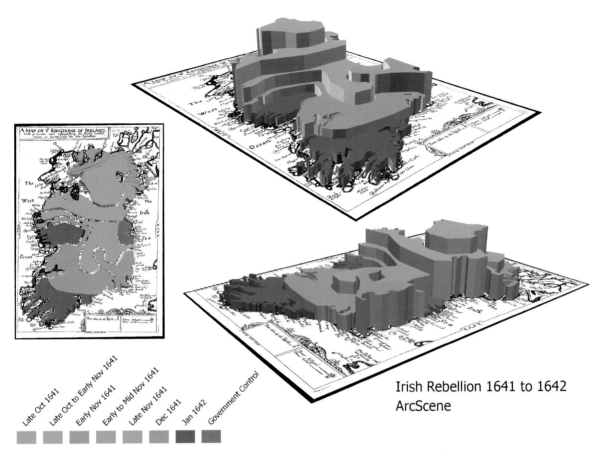

Late Oct 1641 • Late Oct to Early Nov 1641 • Early Nov 1641 • Early to Mid Nov 1641 • Late Nov 1641 • Dec 1641 • Jan 1642 • Government Control

Irish Rebellion 1641 to 1642
ArcScene

Figure 3.2   **ArcScene, Irish Rebellion 1641 to 1642.** Created by the author from *A Map of Ye Kingdome of Ireland* dated to 1655, from the British Library. Author unknown.

maps. A close examination using the functions of the ArcScene platform reveals that a few of the campaign's polylines intersected natural features such as mountain ranges, which may have posed physical obstacles to Cromwell's tactical movements. Examining the terrain of battles and sieges in a 3D mapping environment makes it possible to better grasp the topographical contingencies that may have aided—as well as hindered—Cromwell's army and the Irish resistance to it. ArcScene provides a dynamic, iterative platform that historians can use to visualize and review data and sources, zoom in and out at different sites, and subsequently refine the trajectories of Cromwell's campaign (as well as other military expeditions and historical events) at finer and more topographically relevant scales. Both figure 3.2 (the 1641 Rebellion) and figure 3.4 (Cromwell's campaign in 1649–50) are visualizations of stills captured from ArcScene cinematically produced animations. The ability to model, visualize, and manipulate spatial scales using gaming and filmic techniques in GIS provides a powerful research tool; used in tandem with qualitative methods and critical theory, it can open research avenues and help probe the multiple perspectives of 3D spatialized histories.

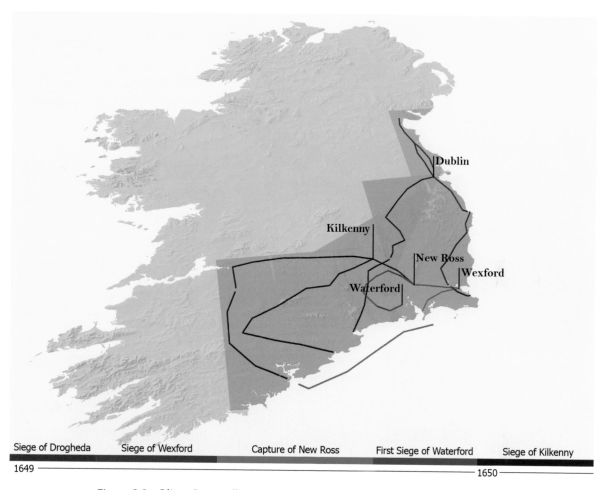

Figure 3.3   Oliver Cromwell's seventeenth-century conquest of Ireland. By the author.

Surveying the Cromwellian Settlement

Private investors financed Cromwell's campaign with funds they secured with a promise of
2.5 million acres of Irish land to be confiscated by the New Model Army. The 1642 Long
Parliament passed the Adventurers' Acts, which failed to meet Cromwell's expenses but
created the legal basis for his creditors to ask for repayment of their guarantees in land. In 1652
Parliament, facing a 1.5 million-sterling deficit due to back pay owed to over 34,000 soldiers,
enacted legislation to provide them with land in lieu of money. Over 7,000 veterans of the New
Model Army eventually settled on confiscated land in what is called the Cromwellian Settlement.
As a consequence, the English transferred substantial acreage from rebellious Catholic owners to
largely Protestant aristocrats, investors, and soldiers. (Dispossessed Catholics of both indigenous
Irish and Anglo-Norman descent were transplanted to Connacht or farther afield.)[16]

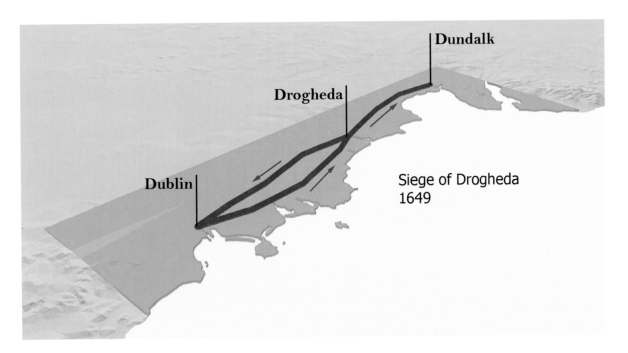

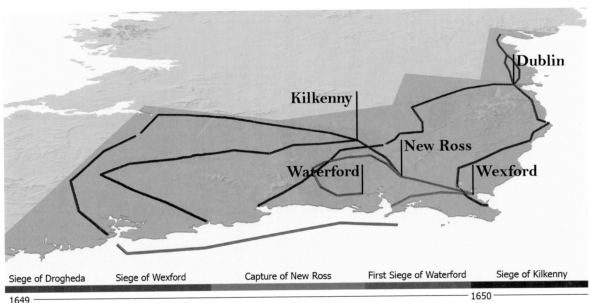

Figure 3.4 **ArcScene topography of Cromwell's campaign in 1649–50.** By the author.

The Cromwellian conquest and its aftermath arguably wielded the most effective hammer in forging the identity of a single Irish-Catholic national identity over the next 400 years.[17] J. S. Wheeler summarizes that this mass redistribution of land profoundly shaped the course of the island's history and demographics of its religious-political geography for centuries to come:

> The land settlement that grew out of the war also transformed Ireland and remains
> one of the defining memories of modern Irish consciousness . . . On the part of Irish
> Catholics, confiscation, transplantation and the slaughter of Drogheda stand as symbols
> of English oppression and Irish suffering. To Irish Protestants, the rebellion of 1641, with
> its attendant horrors, remains a potent element in their fear of Catholic rule. These are
> powerful and depressing legacies of the long Irish war of 1641–52.[18]

The conquerors needed to survey their confiscated land before they could redistribute it. English Surveyor-General Benjamin Worsley conducted one in 1653 that produced flawed results. Sir William Petty (1623–87), the physician-general in Cromwell's army, pointed out its defects and suggested methodological improvements. Political intrigue ensued, but in 1654, Petty successfully undercut the surveyor-general's budget and secured a commission to map the seized lands. He completed his survey in 13 months. On examining its results, Worsley advised its rejection; however, Petty's Down Survey won parliamentary committee acceptance in May 1656.

William Petty and the Down Survey

The Down Survey, also called the Civil Survey, earned its name from Petty because of the "admeasurements" he set "down" in maps. Supplemented by the Strafford Survey of Connaught (or Connacht) of 1636–40 and other surveys conducted from 1656 to 1659, Petty's maps were published as the first modern atlas of Ireland, titled *Hiberniae Delineatio* (1685) (figure 3.5). In 1672, Petty completed a *Political Anatomy of Ireland*, published posthumously in 1691, which drew on his topographical study of the island.[19] Petty orchestrated the Down Survey from his headquarters in Dublin and enlisted and dispatched 1,000 men, many of them soldiers demobilized from Cromwell's army, into the field to take measurements and record topographical features. These men, largely unemployed after the Irish war, were not skilled surveyors. However, Petty laid out their tasks in simple detail, designing and constructing simple tools that enabled them to observe and collect the data he specifically wanted. Petty's surveyors had only to record locations and measure distances by chain lengths he provided. They sent the collected information to the Central Office in Dublin, where Petty supervised a group of trained cartographers who collated and drafted the data onto grid paper. In this sense, he instituted a seventeenth-century "geographical information system" of Ireland, which because of its structure and attention to topographical detail, can be imported, without much alteration, into the digital platform of a twenty-first-century GIS.

Figure 3.5   *Hiberniae Delineatio* (1685). Courtesy of Trinity College Dublin, *Hiberniae Delineatio* (1685).

Using the data he collected from his surveyors, Petty helped draft all the parish maps produced by the Down Survey. The unit of scale he employed was 40 Irish perches to the inch (an Irish perch equaling 21 feet). Later, he edited these maps into barony maps and included terriers (registers of landed property) listing the landowners during the 1640s alongside the new owners, their religious affiliation, land valuation, and area.

From the *ballybetagh* to the barony

Prior to the plantations of Ulster and Munster and Cromwell's conquest of Ireland, Gaelic
territorial, social, and genealogical systems were based on an area measure known in Ulster as
the *ballybetagh* (the three other Irish provinces possessed similar Gaelic denoted territorial
arrangements). The Gaelic territorial clan networks that sustained the ballybetagh, (as they
were called in Ulster) shown in figure 3.6, are merely speculative and intended to only illustrate

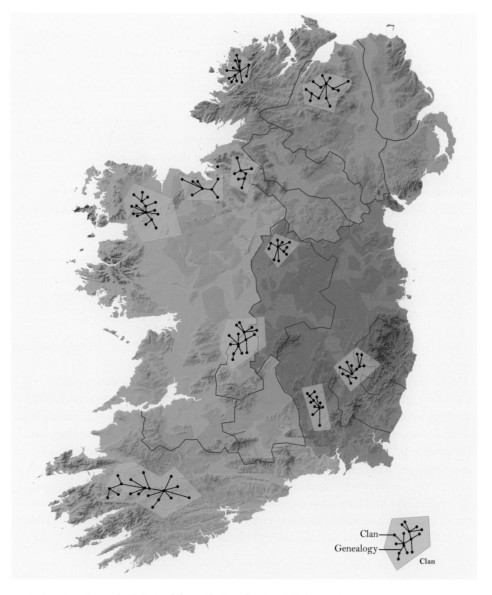

Figure 3.6   **Speculative ballybetagh boundaries of Ireland.** By the author.

possible templates for visualizing the social webs that comprise the fluid nature of tribal historical geographies. Further work needs to be undertaken in GIS to map and visualize actual Gaelic clan locations, genealogies, identities, hierarchies and webs.[20]

Under Gaelic systems, dominant clans expanded at the expense of weaker kin-groups in a well-defined and well-established territorial lattice. For instance, in the Ulster Gaelic system, the ballybetagh, parish, and *ballyboe* (townland) boundaries reinforced each other, generally reflecting the pastoral society of the native Irish. Despite having an extremely low population density by 1600, clans across Ireland participated in large-scale summer migrations to seek pastures for their livestock.[21]

A ballybetagh generally consisted of twelve ballyboes, named to reflect a prominent natural or built feature of the landscape and the names of families who lived on the land. Topographical features, such as *tullagh* (hills), and flora, such as *annagh* (marshes) or *carraig* (rocks), combined with family lore, customs, and genealogy to give specific ballyboes their place-names.[22] Consequently, at the higher ballybetagh level, as land shifted between different family branches, the names of ballybetaghs changed with them.[23] Gaelic society recorded these name changes in an oral geography called the *dindshenchas* (the lore of places). Often, elite poets called *fili* in Gaelic, adjudicated by *brehon* (judges), communicated this information in the form of storytelling. In this way, poetic memory preserved the variable meanings of some of the great family and place-names across Ireland.[24] However, from the sixteenth century onward barony and county boundaries began to replace the ballybetagh as the dominant territorial unit, reinforcing the increasing English political-economic power over the island and its people. As a consequence,

> On a strategic territorial level the English-inspired "barony" [. . .] became a place which occupied the life and knowledge of local ruling elites in the mid-seventeenth century—the barony became central in terms of military mobilization, food supply and local governance.[25]

Thus, Cromwell's campaign and subsequent redistribution of Irish land changed the manner of organizing and naming territories in Ireland.

Unlike his parish maps, Petty's barony maps included townland boundaries and, in some instances, the sites of houses, castles, fields, and roads. The maps also anglicized Gaelic descriptions of natural terrain, listing water features, meadows, bogs, woodlands, mountains, and several types of fields (arable, nonarable, and pasturage). The Down Survey did not include land mapped by the Strafford Survey, which included the Connacht province counties of Clare, Galway, Mayo, and Roscommon in center-west Ireland, where Cromwell had exiled dispossessed Catholics. Petty's survey contributed to the creation of the *Books of Survey and Distribution,* which tabulated the land ownership transfers of the Cromwellian Settlement. The information contained in the *Books* provides a means to create a geodatabase and visualize the redistribution of property and power in seventeenth-century Ireland.

The *Books of Survey and Distribution*

Created between 1676 and 1678, the *Books of Survey and Distribution* comprise 22 volumes, assembled on a territorial basis by county. Incorporating data from the Down Survey, Strafford Survey, and Civil Surveys, the books list the names of landowners compelled to forfeit their holdings under the Cromwellian Settlement, a description of the amount and quality of their land, and the names of those who received their land under Parliament's land redistribution acts passed between 1662 and 1703. This helped the government impose and collect a yearly "quit rent" on landowners.

The index organizes each volume according to an "alphabet" of land denomination, using the barony as the primary geographical unit of record. The compilers probably adopted the barony boundaries from a few primary sources, including the "Books of Reference" for Petty's Down Survey maps and Civil Survey of 1654–56, both of which Petty orchestrated, as well as the earlier Strafford Survey of 1636–40 that covers Connacht in the center-west.[26] Parish subdivisions provide localized information. Each volume identifies county, barony, and parish in bold calligraphic script written across the top of each page. Land transfers are outlined in tabular form on each page. For example, the County Mayo volume enumerates and details accordingly:

"Numbers in the Plot";
"Proprietor's Name [Anno 1641]";
"Denominations of Land";
"Lands Unprofitable"—typically bogs, woods, barren mountains, and loughs;
"Lands Profitable"—typically meadow, arable, and profitable pasture;
"Acres disposed on the Acts"; and
"To whom soe disposed."[27]

This form of tabular outline appears relatively consistent throughout the 22 volumes of the *Books*.

[Left hand side]	[Right hand side]
(1) "Number of reference in the Alphabet	(7) No. of Profitable Acres disposed of on ye Acts
(2) Number of Plot in the Downe Survey	(8) To whom soe disposed with their Tytle whether by Decree, Certificate or Patent Reference to ye Record Thereof
(3) Proprietors in Anno 1640 and their Qualifications	
(4) Denominacons	(9) No. of the Book or Roll and No. of ye Page or Skin
(5) Number of Unprofitable Acres by the Downe Survey	(10) No. of Prof. acres remaining undisposed
(6) No. of Acres profitable by the Downe Survey	

Figure 3.7 **Ledger, *Books of Survey and Distribution*.** Created by the author using data derived from the *Books of Survey and Distribution*, 1636–1703, vols. i, ii, iv; ed. Robert C. Simington (1947, 1956, 1967); and the *Books of Survey and Distribution*, 1636–1703, vol. iii, ed. Breandán Mac Giolla Choille (1962), The Irish Manuscript Commission.

A series of databases using information from the *Books* was created to map the religious demographic transformation of Irish land ownership in the wake of Cromwell's invasion and Parliament's redistribution acts. Here, the barony functions as the unit of scale to illustrate the seventeenth-century land transfer from Catholics to Protestants and distinguish between total, profitable, and unprofitable acreage. Ian Gregory contends that we should regard GIS as a database visualizer.[28] The *Books* contain geospatial and attribute data, which a GIS can format and edit in a digital database and then join to a barony map layer. In this way, we can visualize the data by reconstructing the perspective of those who assembled the volumes. Thus, GIS database and mapping techniques help build a forensic reconstruction of Petty's surveys and the *Books* to analyze the historical and geographical variables of Ireland's violent land transfers.

Database mapping the *Books*

To this end, I collated and parsed information from the *Books*, then reconstructed it in a digital database (Excel, Dbase, Oracle, and Microsoft Access can serve as platforms). I then linked the database, coded with map information, to a barony shapefile layer to create a series of choropleth maps illustrating landownership changes in Ireland between 1641 and 1670.[29] Two graduated symbol maps respectively illustrate the distribution of lands held by the largest Catholics and Protestant landowners in 1641 and 1670.

The map in figure 3.8A, *1641 Top Landowners*, exhibits a somewhat uniform land distribution, leaning slightly toward higher Protestant ownership, perhaps because of English Crown policies from 1603. By comparison, figure 3.8B, *1670 Top Landowners,* clearly shows the effects of the Cromwellian Settlement and the Acts that redistributed land. The map verifies the massive land transfers to the largely Protestant group of investors, settlers, and soldiers loyal to the Crown, except in Connacht where Catholic ownership prevailed. Comparing the two maps, it is interesting to note the continuation of large Catholic landownership in counties Cork in the south and Antrim in the north. The latter is counterintuitive in regards to twentieth-century Irish nationalist tropes, particularly concerning the religious demography of landownership in Ulster. By collating this ledger data in a GIS, which to date had been enumerated in tabular form, we can offer a visual empirical perspective to debates about the political, economic, and religious history of seventeenth-century Ireland.[30]

Another series of four maps created with GIS database-mapping methodologies and techniques break down the religious demography of landownership and illustrate the contiguous presence of Catholics and Protestants in some regions, even at the barony level.

In figure 3.9A, the map *1641 Top Catholic Landowners* shows the locations of Irish members of the aristocracy who were the largest Catholic landholders that year. Indeed, the map exhibits their baronial presence in all four provinces of Ireland. Figure 3.9B, *1641 Top Protestant Landowners*, illustrates a preponderance of Protestant aristocrats holding land in a belt that extends from the Leinster province in the southeast, extends west, and abuts the Munster Plantation (which included Cork) in the southwest.

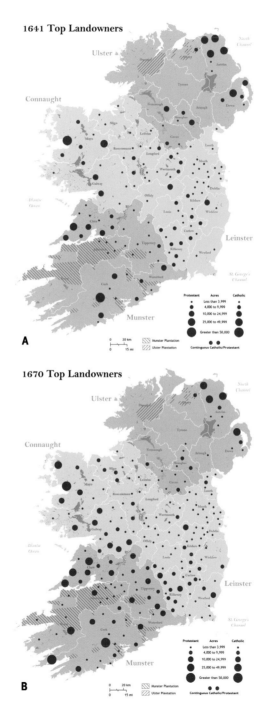

Figure 3.8 **Top landowners, 1641 (A) and 1670 (B).** Created by the author using data derived from the *Books of Survey and Distribution*, 1636–1703, vols. i, ii, iv; ed. Robert C. Simington (1947, 1956, 1967); and the *Books of Survey and Distribution*, 1636–1703, vol. iii, ed. Breandán Mac Giolla Choille (1962), The Irish Manuscript Commission.

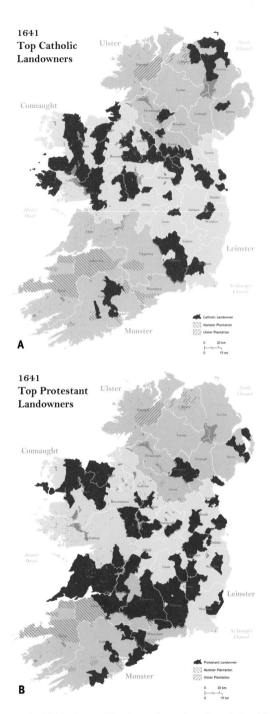

Figure 3.9 **Top landowners in 1641.** Created by the author using data derived from the *Books of Survey and Distribution*, 1636–1703, vols. i, ii, iv; ed. Robert C. Simington (1947, 1956, 1967); and the *Books of Survey and Distribution*, 1636–1703, vol. iii, ed. Breandán Mac Giolla Choille (1962), The Irish Manuscript Commission.

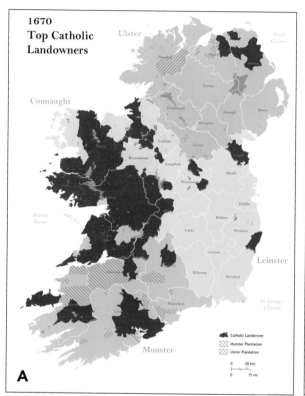
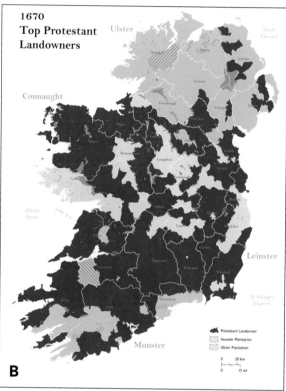

Figure 3.10 **Top landowners in 1670.** Created by the author using data derived from the *Books of Survey and Distribution, 1636–1703*, vols. i, ii, iv; ed. Robert C. Simington (1947, 1956, 1967); and the *Books of Survey and Distribution, 1636–1703*, vol. iii, ed. Breandán Mac Giolla Choille (1962), The Irish Manuscript Commission.

The map *1670 Top Catholic Landowners* in figure 3.10A indicates that aristocratic Catholics were pushed largely into Connacht because of the Cromwellian Settlement and Adventurers' Acts. Significant Catholic strongholds exist in Ulster and Cork provinces (north and south), while Catholic settlement in Leinster province in the east appears largely absent. The map in figure 3.10B, *1670 Top Protestant Landowners*, highlights a mass proliferation of Protestant ownership into an area that extends from the coast of Leinster in the southeast to the Atlantic coast on the west and throughout Munster Plantation (which included Cork) in the south. Protestant landownership predominated in northwest Connacht. A comparison of the four maps also reveals that Catholic and Protestant landowners existed side by side in several baronies. In 1641, Catholics and Protestants both clustered in the provinces of Leinster, around county Kilkenny, and Connacht, around North Mayo and into Roscommon. By 1670, Catholics who were pushed west settled with Protestant landholders (both pre- and post-Cromwell) in the counties of Mayo, Galway, the Atlantic fringes of counties Clare and Kerry, and around Cork city. Importing the barony shapefiles from both top landholder maps into ArcScene makes it possible to visualize the data joins "geostatistically" in 3D. After importing the shapefiles, I extruded the

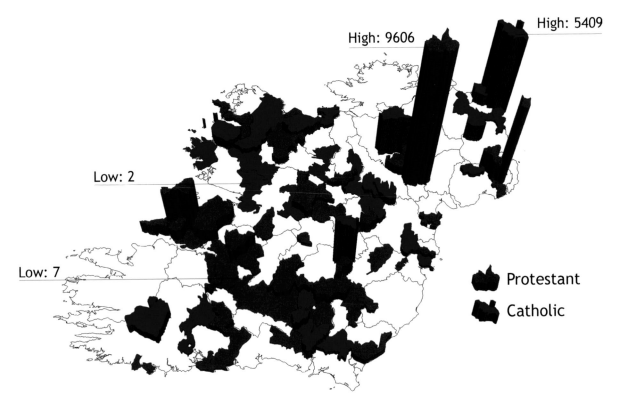

High: 9606

High: 5409

Low: 2

Low: 7

Protestant

Catholic

Figure 3.11 **Top Profitable Acres in 1641.** Created by the author using data derived from the *Books of Survey and Distribution*, 1636–1703, vols. i, ii, iv; ed. Robert C. Simington (1947, 1956, 1967); and the *Books of Survey and Distribution*, 1636–1703, vol. iii, ed. Breandán Mac Giolla Choille (1962), The Irish Manuscript Commission.

variable designating "profitable acres" to illustrate by height the amount of profitable baronial acres held by Catholics and Protestants in 1641 and 1670.

The map *Top Profitable Acres in 1641* (figure 3.11) illustrates that landholdings in Ulster province yielded the highest level of profitability in Ireland for Catholic and Protestant owners alike.

The map *1670 Top Profitable Acres*, (figure 3.12), in keeping with the trends indicated by *1670 Top Landowners* and *1670 Top Protestant Landowners*, indicates that Protestants held a collective majority of profitable acres across the east, south, and west of Ireland. However, the map data also demonstrate that the most profitable acres under single ownership in 1670 belonged to Catholic aristocrats.

The preceding maps, reconstructed from the *Books of Survey and Distribution* illuminate how GIS database techniques can forensically employ archival sources to visualize socio-political and economic trends as they occurred in historical environments.

A fire in the Surveyor-General's Office destroyed Petty's original Down Survey maps in 1711. Two hundred years later, the destruction of the Public Records Office of the Four Courts in Dublin during the Irish Civil War (1922–23) incinerated copies of the maps. Such events

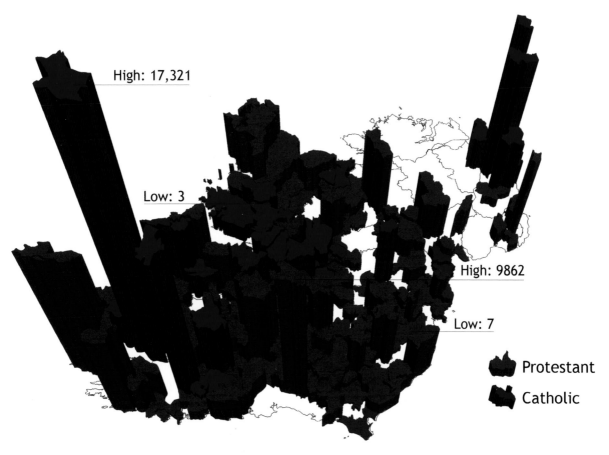

High: 17,321

Low: 3

High: 9862

Low: 7

Protestant

Catholic

Figure 3.12 Top Profitable Acres in 1670. Created by the author using data derived from the *Books of Survey and Distribution*, 1636–1703, vols. i, ii, iv; ed. Robert C. Simington (1947, 1956, 1967); and the *Books of Survey and Distribution*, 1636–1703, vol. iii, ed. Breandán Mac Giolla Choille (1962), The Irish Manuscript Commission.

underscore the relevance of GIS forensic historical techniques. However, various copies of Petty's maps do exist and coupled with data from the *Books* provide the means to reconstruct seventeenth-century cartographical perspectives in GIS.

Visualizing the webs of history

Ian Gregory and Paul Ell contend that historical research typically involves gathering information from a number of different sources and collating it for the purpose of gaining a better understanding of the phenomena being studied. Because most information has a spatial reference, GIS offers the potential to collect and store qualitative and quantitative data types for historical research in ways that would previously have been impractical or even impossible.[31] GIS can also provide a tool of historical inquiry as well as a data-rich secondary source of visual information, through which historians can amplify and communicate their findings to an audience.[32]

This chapter exhibits the utility of GIS for historical study by reconstructing, mapping, and geostatistically visualizing data related to the political, economic, and religious conflicts of seventeenth-century Ireland. This demonstrates how GIS can help historians display and explore data through the creation of spatially coordinated presentation of graphics, thematic maps, and charts.[33] By no means authoritative, this act of mapping forensic data extracted from archival records of the period meets two significant methodological considerations for those endeavoring to develop historical GIS. First, scholars must source, parse, structure, and edit primary data for use in a GIS. Second, they must provide topographical verification of the visualizations created. With the twenty-first century digital revolution and emerging GIS techniques, it is now possible to reintegrate historical and geographical methodologies in dynamic, visual, and cinematic ways only imagined less than a century ago. In 1939, the geographer Richard Hartshorne reflected,

> Theoretically one might construct an unlimited number of separate historical geographies of any region, and if these could be compared in rapid succession one would have a motion picture of the geography of an area from the earliest times to the present.[34]

Hartshorne concluded that it was practically impossible to synthesize history and geography in such a manner.[35] The noted historical geographer H. C. Darby also employed a cinematic metaphor when he stated in 1953 that "the landscape is not a static arrangement of objects. It has become what it is, and it is usually in the process of becoming something different. A close analogy is to regard our momentary glimpse of it as a 'still' taken out of a long film."[36] Now, Hartshorne's and Darby's musings can be realized by incorporating the next generation of GIS presentation graphics that use animation, facilitate interactive data landscapes, and render fully 3D-layered information spaces. Furthermore, by reconstructing archival records using digital GIS database techniques, historical researchers and students can visually query and navigate through complex information spaces to locate and retrieve task-relevant subsets of data.[37]

Pedagogically, history students who engage GIS learn to appreciate how the selection and visualization of data constitute a subjective process and a form of social construction embedded in power relations that often elide, subjugate, and silence historical voices, actors, and places. This certainly was true for the Down Survey and the *Books of Survey and Distribution*. It also underscores John Pickles' observation that:

> One of the most difficult lessons for anyone to learn is the way in which their own worlds are geographically coded; to understand the relationship between the visible and invisible, the proximate and the distant, and to recognize the complex folds of past and present that constitute place and experience as we know it.[38]

Last, by providing highly interactive operations with the capacity to simulate change, GIS combines visual insights with a capability to probe, drill down, filter, and manipulate the digital

display to answer the "why" as well as "what" questions, in addition to the most fundamental
question—where?[39]

Sources

1   J. L. Gaddis, *The Landscape of History: How Historians Map the Past* (Oxford: Oxford University Press, 2002), 32.

2   Ibid., 32, 33–34.

3   J. Smith, *A Generall Historie of Virginia* (Bedford, MA: Applewood Books, 2006), 331.

4   R. Johnston et al., *The Dictionary of Human Geography*, 4th ed. (Malden: Blackwell, 2000), 410.

5   D. Massey, *For Space* (London: Sage, 2005); Merriman et al., "Space and Spatiality in Theory," 9; M. F. Goodchild, "Time, Space, and GIS," *Past Place: The Newsletter of the Historical Geography Specialty Group,* 14, no. 2 (2006): 8–9; T. Hägerstrand, "Space, Time and Human Conditions" in *Dynamic Allocation of Urban Space*, eds. A. Karlqvist, L. Lundqvist, and F. Snickars (Lexington, MA: Lexington Books, 1975), 3–14; D. G. Janelle, "Time-Space in Geography," in *International Encyclopedia of the Social and Behavioral Sciences*, eds. N. J. Smelser and P. B. Baltes (Amsterdam: Pergamon-Elsevier Science, 2006), 15,746–49; M.-P. Kwan and T. Schwanen, "Critical Space-Time Geographies," *Environment and Planning A,* 44, no. 9 (2012): 2,043–48; M. J. Kraak, "Geovisualization Illustrated," *ISPRS Journal of Photogrammetry and Remote Sensing,* 57 (2003): 390–99; A. Pred, ed., *Space and Time in Geography: Essays Dedicated to Torston Hagerstrand* (Lund: Gleerup, 1981); N. Thrift and J. May , eds., *Timespace: Geographies of Temporality* (London and New York: Routledge, 2001); I. Gregory and P. Ell, *Historical GIS: Technologies, Methodologies and Scholarship* (Cambridge: Cambridge University Press, 2007).

6   P. J. Ethington, "Placing the Past: 'Groundwork' for a Spatial Theory of History," *Rethinking History,* 11, no. 4 (2007): 465–93, 487; B. Fay, "Environmental History: Nature at Work," *History and Theory*, 42, no. 4 (2003): 4.

7   Ethington, "Placing the Past," 487.

8   Ibid., 466.

9   Ibid., 466, 487.

10 J. B. "Jack" Owens, "What Historians Want from GIS," *ArcNews Online*, Summer 2000, accessed April 23, 2012, http://www.esri.com/news/arcnews/summer07articles/what-historians-want.html.

11 A. K. Knowles, *Placing History: How Maps, Spatial Data, and GIS Are Changing Historical Scholarship* (Redlands: Esri Press, 2008), 3, 10.

12 M. Dodge, M. McDerby, and M. Turner, "The Power of Geographical Visualizations," in *Geographic Visualization: Concepts, Tools and Applications* (Chichester, England: Wiley, 2008), 3–4.

13 P. Lenihan*, Confederate Catholics at War 1641–1649* (Cork: Cork University Press, 2001), 12.

14 William Web, *A Map of Ye Kingdome of Ireland* (Oxford; Webb, 1642 [1655]).

15 Pattern sourced from J. S. Wheeler, *Cromwell in Ireland* (New York: St. Martin's Press, 1999).

16 F. O'Toole, "*Books of Survey and Distribution*, Mid-17th Century," *Irish Times,* May 5, 2012, http://www.irishtimes.com/culture/art-and-design/books-of-survey-and-distribution-mid-17th-century-1.515490.

17 Smyth, *Map-Making, Landscapes and Memory*, 149.

18 Wheeler, *Cromwell in Ireland*, 227.

19 Sir William Petty, *Hiberniae Delinatio* (Shannon: Irish University Press, 1969 [1685]); Sir William Petty, *The Political Anatomy of Ireland* (Sir William Petty's political survey of Ireland) (London; Browne, Mears, Clay and Hooke, 1719 [1691]).

20 Smyth, *Mapmaking, Landscapes and Memory*, figure 3.9, p. 78; and figure 3.10, p. 81.

21 Smyth, *Mapmaking, Landscapes and Memory*, 73–75; R. Foster, *Modern Ireland: 1600–1972* (London: Allen Lane–Penguin Press, 1988), 16.

22 Public Record Office of Northern Ireland, "The Townland," Local History Series: 1 (2007).

23 Smyth, *Map-Making, Landscapes and Memory*, 73–75.

24 Ibid.

25 Ibid. 146–47.

26 R. C. Simington, ed., *Books of Survey and Distribution*, vol. 2, *County of Mayo* (Dublin: Stationary Office, 1956), http://www.irishmanuscripts.ie/servlet/Controller?action=digitisation_backlist.

27 Ibid.

28 I. Gregory, lecture for workshop, "Geospatial Methods for Humanities Research," Digital Humanities Observatory Summer School, Royal Irish Academy, Trinity College, summer 2010.

29 Template maps originally drafted by the author for J. Ohlmeyer, *Making Ireland English: The Irish Aristocracy in the Seventeenth Century* (New Haven: Yale University Press, 2012); data originally sourced from Simington, ed., *Books of Survey and Distribution*.

30 The *Books of Survey and Distribution* data in this chapter was visualized by choropleth (translated from the Greek 'area/region' and 'multitude') map schemas, in which regions are displayed typically by administrative units (such as the barony) that are colored or symbolized in proportion to the statistical variable being displayed. It is suggested that further studies utilizing dasymetric mapping techniques with parish and townland shapefile layers would more accurately distribute attribute data within or across the baronial administrative units by the overlay of geographic boundaries that exclude, restrict, or confine the attribute data in question to its specific location. For example, landowner religious attributes organized by the baronial unit of measurement might be more accurately distributed by the overlay of landscape features, such as bogs, water features, mountains, where it is reasonable to infer less population, in addition to aggregating such data to parish and townland layers to more finely scale the representation of land distribution demographics. Dasymetric mapping can limit the error of "ecological fallacy" in statistical interpretation and representation. Ecological fallacies are inferences made about the nature of individuals or the features of a region based solely on aggregate statistics collected for the group or region to which those individuals belong or are located.

31 Gregory and Ell, *Historical GIS*, 37.

32 D. J. Staley, *Computers, Visualization and History* (Armonk: Sharpe, 2003), 125.

33 Dodge, et al., *Geographic Visualization*, 3–4.

34 R. Hartshorne, *The Nature of Geography: A Critical Survey of Current Thought in the Light of the Past* (Lancaster, PA: The Association Lancaster, 1939), 138.

35 Ibid.

36 H. C. Darby, "On the Relations of Geography and History," *Transactions and Papers of the Institute of British Geographers,* 19 (1953): 7.

37 Ibid.

38 Pickles, *A History of Spaces*, 81.

39 Dodge et al., "The Power of Geographical Visualizations," 3–4.

Part 2

Writers, texts, and mapping

Chapter 4

GIS and the poetic eye

Mapping Kavanagh

The German literary critic Walter Benjamin once reflected that for years he had "played with the idea of setting out the sphere of life—*bios*—graphically on a map."[1] Humanities GIS techniques make it possible to chart the trajectories of a writer's biography and plot the resulting spaces alongside the distinct historical poetics of place that emerge from the writer's successive works. This type of GIS mapping, according to David Staley, allows "a kind of multidimensional emplotment: a single story organized from multiple and heterogeneous elements," which can imbue "a spatial totality" on the study of a particular writer's experience and representation of place as it evolves over time.[2]

Engaging GIS to map such a critical literary geography intimates the splendid ambiguity of a journey into the unknown. The County Monaghan poet Patrick Kavanagh (1904–67) undertook a similarly nomadic trek when he embarked from his rural parish of Inniskeen, nestled in the hilly landscape of South Ulster, to Dublin, the capital of *Saorstát na nÉireann* (Irish Free State), in 1931. Of this experience, he wrote,

> Ten miles from home I was in strange country, among folk who wouldn't know me [...]
> A granite milestone along the way told me it was forty-five miles to Dublin: the chiselled
> letter-grooves were filled with moss and I had to trace the letters with my finger like a
> blind man reading Braille.[3]

Here, GIS served to model and juxtapose the patterns of Kavanagh's daily life as a farmer-poet in the 1920s and 1930s with patterns he established after relocating to Dublin in 1939. It helped to distinguish the differences between his rural and urban patterns that may have influenced a shift in his poetic perception and style. Memories, Gaston Bachelard writes, "are motionless, and the more securely they are fixed in space, the sounder they are."[4] Born in a small cabin located slightly south-west of Inniskeen village, on the rise of a small drumlin, Kavanagh's earliest memory was anchored in the hill-scape of his native townland:

> The name of my birthplace was Mucker [...] the name was a corrupted Gaelic word
> signifying a place where pigs bred in abundance.[5]

Given the location of his family's cabin, Kavanagh writes, "The western sun, without regard for the laws of men, peeped through our small back window."[6] His prose helps us to further imagine the environs of his birthplace: "The railway line to Carrick was visible from our back window. One hundred yards from our house was the railway gate-house and level crossing where the Mucker Road joined the country highway."[7] The son of a cobbler and small farmer, Kavanagh began writing poetry in his father's fields, which he was set to inherit. He began composing verses in the 1920s, drawing inspiration from his rural surroundings. On the eve of the Second World War, he left the land to pursue life as a writer in the city of Dublin. Kavanagh's life and literary work embody the cultural and demographic shift that set the course of twentieth-century Ireland as migration to the island's urban centers drained the countryside of its population. Kavanagh's poetry and prose depict the rural landscapes, fields, townlands, and parish in which he was raised as well his urban impressions of the city, with its Georgian streets and canals. He died in 1969, an established but impoverished writer in Dublin.

In the late summer of 2007, I traveled to the Kavanagh Visitor's Center nestled in the rounded drumlin fields of South Ulster. The doyen of Irish geography, E. Estyn Evans, states that the majority of Irish place-names possessing *druim,* or drum (which means "hill" in Gaelic), occur in this region. Evans notes that a confusion of little hills, winding streams, small lakes, and bog lands make up this terrain.[8] At the regional center, housed in a deconsecrated stone church off the main village road of Inniskeen, I witnessed Kavanagh's cinematic resurrection. The ghostly light of a film projector conjured Kavanagh back to life, like an ancient Irish *fili* (poet), as rain pounded the slate roof of the church. The poet's flickering face glowed like a specter on the movie screen, his thick-lensed, horn-rimmed glasses perched upon a prominent nose, giving him a magisterial, if detached, air. Reciting his poetry, Kavanagh seemed to float above the stone floor of the church, a magpie's nest of white hair circling his bald pate. Outside, in the grey Ulster rain, his gravestone stood like a broken rune in the lush, green grass of the parish cemetery. Inside, the grainy black-and-white documentary film spooled onward, with Kavanagh incanting with a voice like wet gravel slipping from a shovel, the rural wisdom of not caring and the freedom that this disposition of mind could provide artists and lay people alike.

In his semi-biographical novel *The Green Fool* (1939), Kavanagh declared that as he wandered about the roads and fields he composed his verses.[9] Sitting in the dark church, I could imagine the embodied performance of the young farmer-poet crossing a drumlin field in the rain, perhaps looking for a lost calf, composing a verse of poetry as he trudged along. Kavanagh's verses carry their own rhythms, and in addition to imparting a strong sense of place, perhaps the cadence of his walks through the hills and fields of Inniskeen Parish created rhythms and patterns that impressed themselves later in the spaces of his writing. In *The Place of Writing* (1989), the Nobel Prize winning poet Seamus Heaney observes,

> The usual assumption, when we speak of writers and place, is that the writer stands in some directly expressive or interpretative relationship to the milieu. He or she becomes a voice of the spirit of the region. The writing is infused with the atmosphere, physical and emotional, or a certain landscape or seascape.[10]

The poet's "resurrection" through the technology and medium of film inspired me to explore, through GIS, the daily patterns Kavanagh created as he moved through his landscape to visualize how their rhythms and cadences may be correlated with the changing perspectives of Inniskeen Parish cast by his poetic eye, voice, and pen.

Bakhtinian GIS

Two distinct senses of place distinguish Kavanagh's depiction of Inniskeen in the collection *Ploughman and Other Poems* (1936) and the novel *The Green Fool* (1939) from its portrayal a few years later in the epic poem *The Great Hunger* (1942). The poems in the collection and rural depictions in the novel, for the most part, are infused with a topophilic and pastoral impression of the people and the drumlins of his native parish. The epic poem, in contrast, casts the same landscape under an "apocalypse of clay."[11] Written from the perspective of an urban milieu, it elicits a sense of topophobia verging on topocide or the annihilation of place.[12] To visualize Kavanagh's *lifepaths* in County Monaghan and Dublin, which respectively correspond with his collection-novel and epic poem, I focused Bakhtin's historical poetics and its primary concept of the chronotope through the lens of a GIS to analyze how Kavanagh's separate rural and urban daily patterns corresponded with changes in his poetic perspective toward Inniskeen Parish.[13] This type of "Bakhtinian GIS" collated lifepath patterns surveyed from fieldwork in the drumlins of County Monaghan and the Georgian streetscapes of south Dublin with data drawn from biographical and literary sources. The chronotope (literally "time-space") serves as the primary point from which scenes in literary texts unfurl in narrative space and is central to Bakhtin's historical poetics, which originates from his evolutionary study of literature, commencing with Greek epics and progressing through the folkloric tales of medieval Europe to the modern novels of Rabelais, Flaubert, Stendhal, Balzac, and Dostoevsky.[14] Bakhtin held that the *living* artistic perception of a writer seizes the chronotope in all its wholeness and fullness, from a defined and concrete location. Such an emplacement serves as the basis for the creative imagination and condenses historical time into space.[15] Through the lens of the chronotope, time thickens, takes on flesh, and becomes artistically visible, while space becomes charged and responsive to the dynamics of plot and history.[16] From the classics, Bakhtin identified the chronotope of the idyll, the road, and the public square: from medieval literature, the carnivalesque and from nineteenth-century literature, the salon, the parlor, and the petty-bourgeois town.[17] Conceptualizing and employing a "Bakhtinian GIS" provides a means to employ the chronotope as an optic for reading literary texts as x-rays of the forces working in the cultural systems from which they spring.[18] GIS enhances the bracketing function of the chronotope to help to visualize what Bakhtin calls the "phenomenology of historical space" by carefully differentiating one historical moment from another.[19]

Within a pair of corresponding 3D GIS "time-space cubes," I could parse, collate, and plot Kavanagh's biographical and literary data from the 1920s, 1930s, and 1940s to differentiate the periods associated with his rural and urban lifepaths. GIS made it possible to visualize how the author's daily performances in both rural landscapes and urban streetscapes shaped the images, metaphors, syntax, and rhythms in his poetry. In Kavanagh's poetic canon, the parish, the townland, the field, and their associated Gaelic place-names act as chronotopic nodes through which his poetic narratives about the rural society and landscape he lived in and wrote about thread together. After relocating to Dublin in 1939, his nomadic performances in the urban spaces of the city took on a different pattern, featuring urban nodes, daily rhythms, and literary perspectives on his native place.

Creating a digital *dinnseanchas*

Kavanagh's poetic voice descended from the Gaelic bardic tradition of *dinnseanchas* (the knowledge of the lore of places).[20] In Inniskeen, his writing style reflected an intimate geography based on *seanchas*, a phenomenological confluence between the general principles of topography and the intimate knowledge of particular places.[21] This sensibility, in which Gaelic place-names both anchor and color the poetic imagery of his 1936 collection, is rooted in Bakhtin's chronotope of the "idyll" denoted by:

> An organic fastening-down, a grafting of life and its events to a place, to a familiar territory,
> […] the conjoining of human life with the life of nature, the unity of their rhythm [and] the
> common language used to describe phenomena of nature and the events of human life.[22]

I created the base of the Inniskeen time-space visualization by digitizing and geo-rectifying two historical maps (nos. 29 and 32) from the 1911 Ordnance Survey (OS) and then joining the images in a single raster layer.

The GIS visualization in figure 4.1 contains the four following sites whose Gaelic place-names were anglicized by the original British OS in 1835:

1. Kavanagh's native townland of Mucker.
2. The fields of Shancoduff (*Seanchua Dubh* "Black Hollow").
3. The fields of Drumnagrella (*Droin* "Ridge").
4. The village of Inniskeen (*Inis Caoin* "Beautiful Island").

In the visualization, three polylines in ascending order illustrate Kavanagh's movements over the course of a day and form a composite lifepath pattern of his early years in Inniskeen Parish. The visualization was created from biographical data supplied and a fieldwork survey conducted in the parish. The latter consisted of walking and driving, visiting sites, and tracking paths and routes that Kavanagh might have taken over the drumlin terrain on a daily basis. Although the routes mapped are speculative, they are also pragmatic in the sense that the drumlin topography of the townlands in the parish has changed very little, if at all. The shortest contemporary pedestrian routes conform to an observation that Kavanagh once made:

> My idea of a cultural parochial entity was the distance a man would walk in a day in any
> direction. The center was usually the place where oneself lived though not always.[23]

Figure 4.1   Patrick Kavanagh's lifepaths, Inniskeen Parish, 1920–39. The accompanying photo shows the author in Inniskeen in the 1960s, surveying the landscape of his past. Created by the author using data derived from the British Ordnance Survey (OS) map of Ireland, layers 29 and 32 (1911), from Trinity College Dublin Library. Photo courtesy the National Library of Ireland, from the Wiltshire Photographic Collection, call number WIL pk2[7], published/created 1963.

The visualization in figure 4.1 features three main polylines illustrating a composite picture of Kavanagh's daily movements:

1. Rising before dawn (violet polyline), Kavanagh would have performed chores around his family's house in Mucker.

2. Then, energized from a full night's rest, he probably set out to work in the farthest four fields of Shancoduff, located on the northern slope and basin of a sizable drumlin.

Fieldwork surveys noted that it took approximately 15 to 20 minutes to walk from Mucker to Shancoduff, a circuitous route (violet polyline), because of the formidable elevation of the drumlin. Kavanagh would have spent different amounts of time working in the fields, depending on the day and season. Perhaps he would eat lunch on site or return to Mucker.

3. Allegedly, in the afternoon (yellow polyline), he generally went to the fields at Drumnagrella, adjacent to his family's holding.

4. His poetry (such as *Inniskeen Road*) suggests that on at least some evenings (fire-red polyline), he traveled to Inniskeen village to socialize.

The three respective polylines illustrating Kavanagh's speculated daytime movements locate his anchor in his family's home at Mucker. In the visualization, a composite pattern of centrality emerges, linking this site by the routes he may have walked to the family's fields and village of Inniskeen. Kavanagh depicts his birthplace and the surrounding townlands in *Inniskeen Road: July Evening* (1935):

> The bicycles go by in twos and threes-
> There's a dance in Billy Brennan's barn to-night,
> And there's the half-talk code of mysteries
> And the wink-and-elbow language of delight
> [...]
> A road, a mile of kingdom. I am king
> Of banks and stones and every blooming thing.[24]

The poem is laced with lyricism, dry humor, and laconic observation. It is suffused with magical realism and a mystic-like illumination of the parish's surrounding vista and its idiosyncratic social geographies. However, *The Great Hunger*, published in 1942, deconstructs the centered spatiality of Kavanagh's earlier life as a farmer-poet and reveals a topophobic vision of Inniskeen Parish:

> He stands in the doorway of his house
> A ragged sculpture of the wind
> October creaks the rotted mattress
> The bedpost fall. No hope. No. No lust.
> The hungry fiend
> Screams the apocalypse of clay
> In every corner of this land.[25]

Juxtaposing the GIS time-space visualizations of Kavanagh's composite rural and urban daily lifepaths (figure 4.2) makes their contrasting spatial patterns clear: the former (figure 4.2A) is centrifugal, and the latter (figure 4.2B) is centripetal. Idyllic images associated with the poems he composed in a rural environment transform, and the chronotope of *peripherality*, signifying a place "lost in a cyclical, natural or static time-warp, forgotten by history, and bypassed by history,"[26] which begins to shape his poetic perception of Inniskeen Parish.

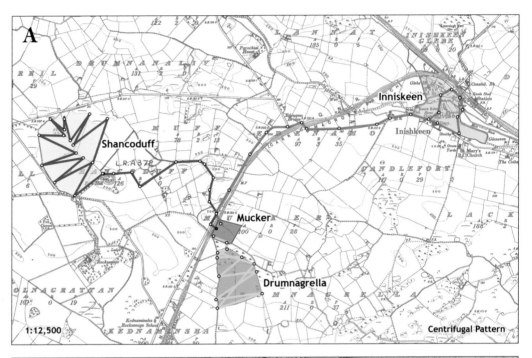

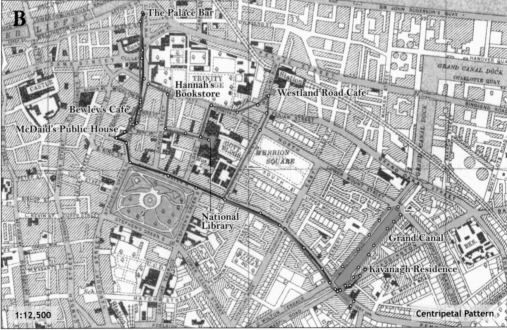

Figure 4.2   **(A) Inniskeen Parish (centrifugal pattern) and (B) Dublin (centripetal pattern).** Created by the author using data derived from the British Ordnance Survey (OS) map of Ireland, layers 29 and 32 (1911), from Trinity College Dublin Library; and Saorstát Eirann Ordnance Survey (OS) Dublin & Environs 1: 20000 Sh 265b.

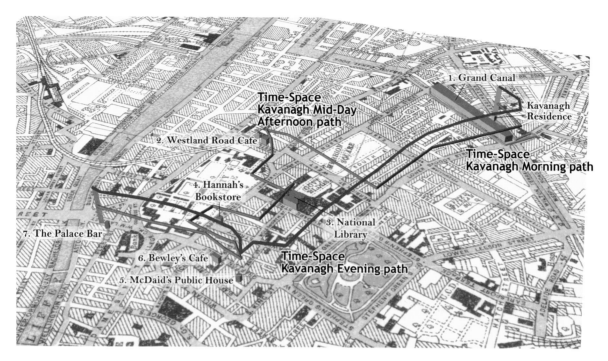

Figure 4.3   Patrick Kavanagh's lifepath, Dublin, 1939–42. Created by the author using data derived from the Saorstát Eireann Ordnance Survey (OS) Dublin & Environs 1: 20000 Sh 265b, published in 1934, from Trinity College Dublin Library.

The GIS visualization in figure 4.3 records the fieldwork exercise that I used to identify the various sites in Dublin that Kavanagh was known to frequent in the early 1940s. I reconstructed his movements through the streets of the city by plotting polylines within the urban time-space cube in an ascending manner:

1.   Socialized as a farmer to rise early, Kavanagh would write in his bedsit and frequent the Grand Canal in the morning. The green polyline in the upper right hand corner of the GIS visualization captures the speculative movements of his morning routine.

2.   In the afternoon, Kavanagh would often walk to meet his brother for lunch at a Westland Road cafe and afterward wander into a "cultural sphere," which contained the National Library on Kildare Street, Hannah's Bookstore on the corner of Dawson and Nassau Streets, and Bewley's Cafe on Grafton Street. The fire-red polyline charts these midday and afternoon movements.

3.   His afternoon and evening perambulations could overlap with stints in McDaid's Public House on Harry Street and the Palace Bar on Fleet Street before walking back to his apartment near the Grand Canal, symbolized by the purple polyline.

Located across Westmoreland Street from the old *Irish Times* offices, the Palace Bar was as "A Café Literaire."[27] Its space approximates Bakhtin's chronotope of the "salon and parlor," where "the webs of intrigue are spun, denouements occur and dialogues finally take place."[28] In Bakhtinian historical poetics, the Palace Bar served as the barometer of political and business life and acted as a space where political, business, social, and literary reputations were made and destroyed, as well as careers begun and wrecked.[29] Bakhtin notes "that the graphically visible markers of biographical and every day time are concentrated, condensed, and focused"[30] by such a space. The Palace Bar provided Kavanagh a crucial entry into Dublin's literary circles, where he secured writing commissions and began to establish his name as a man of letters. Nicknaming it the "Malice Bar," Kavanagh recalled that Bertie Smyllie, "the giant Hemingway-esque editor of the *Irish Times*," instituted a nightly branch office of the newspaper at the bar.[31] Antoinette Quinn, a chronicler of the scene, observed, "Almost everyone who counted in journalism and the arts was to be seen in the Palace Bar at some time on the evening of the week: F. R. Higgins, poet and Abbey Theatre director; and M. J. MacManus, novelist, and literary editor of the Irish Press."[32] One can juxtapose the vibrancy of the bar against the insularity of the rural pub, as depicted in *The Great Hunger*:

> The frosted townland of the night.
> Eleven o'clock and still the game
> Goes on and the players seem to be
> Drunk in an Orient opium den.[33]

After the Palace Bar closing time of 11 p.m., Kavanagh would return to the small cramped apartment that he shared with his brother (the red polyline). By employing GIS techniques in conjunction with Bakhtin's historical poetics, Kavanagh's literary and biographical landscapes become "not only graphically visible in space, but also narratively visible in time."[34] Kavanagh's spatial performances in Inniskeen Parish (figure 4.1) convey such a perspective, as captured in his poem, *Ploughman* (1930):

> I turn the lea-green down
> Gaily now,
> And paint the meadow brown
> With my plough.[35]

When juxtaposed against his streetscape perambulations in Dublin (figure 4.3), this pattern reverts to a cyclical spatiality "bound" to an insular, closed social system, as represented in *The Great Hunger*:

> But the peasant in his little acres is tied
> To a mother's womb by the wind-toughened navel-cord
> Like a goat tethered to a stump of a tree—
> He circles around and around wondering why it should be.[36]

The rural landscape in the GIS time-space visualizations of Inniskeen Parish imply centrifugal movements in space, which appear rooted in a locus, yet remain fixed in place, stagnating and ultimately leading nowhere. Perhaps the spontaneity of urban space provided Kavanagh a place to reflect on the social realities plaguing Inniskeen. Kavanagh intimated as much in *The Green Fool* (1938):

> When I arrived in Mucker the natives were beginning to lose faith in the old, beautiful things. The ghost of a culture haunted the snub-nosed hills.[37]

After the Famine of 1847, "keeping the family name on the land, extending farm size if possible, and protecting the holdings from encroaching or jealous neighbors was a local preoccupation" in Kavanagh's region of South Ulster.[38] Large families occupied comparatively small farms due to the failure of older unmarried brothers and sisters to leave the parental home. The practice of primogeniture (in which the eldest son or daughter inherits the family farm) made marriage impossible for many siblings from large families. This phenomenon grew in many small South Ulster farm communities in the 1930s, '40s, and '50s.[39] *The Great Hunger* countered the conventional wisdom of nationalist critics and state leaders who supported a nostalgic "Gaelic Eden."[40] In 1943, the Irish Prime Minister Eamon DeValera delivered a St. Patrick's Day address on Radio Éireann, strikingly blind to the social realities depicted in Kavanagh's poem. His aspirations for Éire envisioned:

> A land whose countryside would be bright with cosy homesteads, whose fields and valleys would be joyous with the sounds of industry, with the romping of sturdy children, the contests of athletic youth, the laughter of happy maidens.[41]

Ironically, the country's 1936 census soon illustrated that southern Ireland contained the highest percentage of unmarried women and men in the world. Related incidence of late marriage, bachelorhood, spinsterhood, and population decline correlated most closely to small farm holdings in regions such as South Ulster and implied the decline, ruin, and entropy of a society. In his epic poem, Kavanagh refuted "the old idea of having your roots in the soil" and observed, "it seems to me it is not a good thing to be stuck to the ground."[42]

In contrast to the "rooted" lifepath pattern established in Inniskeen, Kavanagh's daily patterns in Dublin were characterized by *synekism*, a term coined by geographer Edward Soja to convey the spatial dynamics, stimuli, and agglomeration that accrete around a dominant and centripetal urban core.[43] GIS visualizations imply that a *synekistic* shift in Kavanagh's spatial movement and environment could have contributed to the chronotopic shift in his literary perception of his native parish and its society. Soja notes that synekism provides a spatial context for the active and affective processes of innovation, development, growth, and change.[44] In Dublin, Kavanagh's centripetal movements led to interactions with dynamic nodes of culture, linkages with different networks of knowledge, and new forms of thought and style. We can see in the GIS visualizations that a core space created by the collective dynamics of his urban pedestrian trajectories (figures 4.2 and 4.3) uprooted his earlier, centralized daily lifepath in Inniskeen, thereby freeing his poetic eye from the low horizons of Mucker to gaze on the kaleidoscopic streetscapes of his new urban milieu.

Plotting the poetic eye

GIS allows one to perceive and represent how Kavanagh's daily lifepath patterns in rural and urban space were *deterritorializing* in relation to each other. From a Foucauldian perspective, the contrasting topophilic/topophobic heterotopias represented in *Ploughman and Other Poems*, *The Green Fool*, and *The Great Hunger* illustrate that Kavanagh's pastoral and apocalyptic chronotopic framings convey the contiguous and overlapping *idyllic* and *peripheral* dimensions of Inniskeen Parish. This becomes evident when the works are spatially parsed, storyboarded, and visualized in a GIS time-space cube. The idiosyncratic geographies performed by Kavanagh on a daily basis reflect the variability and unpredictable agency of his lifepath and how it may have shaped the textual spaces and rhythm of his poetry. Such a particularly spatial awareness informs Kavanagh's poem, *I Had A Future* (1952), a bittersweet reflection on his early days in Dublin:

> Gods of the imagination bring back to life
> The personality of those streets,
> Not any streets
> But the streets of nineteen forty.
> [...]
> It is summer and the eerie beat
> Of madness in Europe trembles the
> Wings of the butterflies along the canal.[45]

As previously noted, Kavanagh's writing carries on the Gaelic bardic tradition of *dinnseanchas*, an intimate blend of geography and storytelling that draws heavily on the lore of place knowledge. Storytelling in GIS, as David Staley observes, ergodic and nonlinear, simultaneously reinforces and disrupts chronological sequences of time and space. Because of the nature of this type of storytelling, GIS provides a platform from which to synchronize layers of images, words, and vectors into contrasting and multidimensional narratives that can illuminate, as John Corrigan puts it (echoing Kavanagh's poem), how "a humanities GIS chases the butterfly rather than nets and pins it."[46] Narrative analysis in GIS can focus on "re-storying" people's individual experiences, according to Mei-Po Kwan, by reorganizing and reinterpreting such experiences in relation to relevant social, cultural, and political contexts. In this manner, GIS can help analyze key elements of people's stories (place, plot, and scene) and their experiences chronologically—or not—by rewriting the indices of such stories in terms of a chosen analytical framework.[47] Although it can be argued that the "Bakhtinian GIS" lens engaged in this chapter netted and pinned Kavanagh's rural and urban lifepaths within the confines of two time-space cubes, these schemas are only a part of an iterative process that emphasizes the dynamism of the verb *mapping* over the static denotation of the noun *map*. Nonetheless, they further establish humanities GIS methodologies as a useful tool for chasing questions down the almost inexhaustible avenues of inquiry crisscrossing at the intersections of biography, literature, place, and time.

Sources

1   W. Benjamin, *One Way Street and Other Writings* (London: New Left Books, 1928), 295.
2   Staley, "Finding Narratives of Time and Space," 43.
3   P. Kavanagh, *The Green Fool* (1938; repr. London: Penguin, 2001), 223–24.
4   G. Bachelard, *The Poetics of Space*, trans. M. Jolas (New York: Orion Press, 1964), 9. On memory and landscapes, see D. Bodenhamer, "Creating a Landscape of Memory," 107.
5   Kavanagh, *The Green Fool*, 8.
6   Ibid., 14.
7   Ibid., 9.
8   E. E. Evans, *The Personality of Ireland: Habitat, Heritage and History* (Dublin: Lilliput Press, 1992 [1973]), 27.
9   Kavanagh, *The Green Fool*, 188.
10 Seamus Heaney, *The Place of Writing* (Atlanta: Scholar's Press, 1989), 20–21.
11 P. Kavanagh, *The Great Hunger* (Dublin: Cuala Press, 1942), 33.
12 J. D. Porteous, "Topocide: The Annihilation of Place," in *Qualitative Methods in Human Geography*, eds. J. Eyles and D. M. Smith (Cambridge: Polity Press, 1988).
13 For a more detailed discussion, see C. Travis, "Geographical Information Systems (GIS) for Literary and Cultural Studies: 'Mapping Kavanagh,'" *International Journal of Humanities and Arts Computing*, 4 (2010): 17–37.
14 M. Bakhtin, *The Dialogic Imagination: Four Essays*, trans. M. Holquist, eds. C. Emerson and M. Holquist (Austin and London: University of Texas Press, 1981), 425–26.
15 M. Bakhtin, *Speech Genres and Other Late Essays*, trans. V. McGee, eds. C. Emerson and M. Holquist (Austin and London: University of Texas Press, 1986), 49; M. Bakhtin, "Forms of Time and of the Chronotope in the Novel: Notes toward a Historical Poetics," in *Narrative Dynamics: Essays on Time, Plot, Closure and Frames*, ed. B. Richardson (Columbus: The Ohio State University Press, 2002), 16.
16 Bakhtin, *The Dialogic Imagination*, 81.
17 Ibid., 250.
18 Ibid., 425–26.
19 K. Hirschkop, "Introduction: Bakhtin and Cultural Theory," in *Bakhtin and Cultural Theory*, eds. K. Hirschkop and D. Sheperd (Manchester: Manchester University Press, 1989), 13.
20 D. Kiberd, *Inventing Ireland*, (London: Verso, 1997), 107.
21 C. Bowen, "A Historical Inventory of the *Dindshenchas*," *Studia Celtica*, 10–11 (1975–76): 115.
22 Bakhtin, *The Dialogic Imagination*, 225.
23 P. Kavanagh, *Patrick Kavanagh: Man and Poet*, ed. P. Kavanagh (Maine: Kavanagh Press, 1986), 243.
24 P. Kavanagh, "Inniskeen Road: July Evening" in *Collected Poems*, ed. P. Kavanagh (New York: The Hand Press, 1972), 19.
25 Kavanagh, *The Great Hunger*, 33.
26 J. Leersson, "The Western Mirage: On Celtic Chronotope in the European Imagination," in *Decoding the Landscape*, ed. T. Collins (Galway: Center for Landscape Studies, 1994), 4.
27 C. Connolly, HORIZON, V, (1942): 36.
28 Bakhtin, *The Dialogic Imagination*, 246.
29 Ibid., 247.
30 Ibid.
31 P. Kavanagh, *Lapped Furrows: Correspondence, 1933–1967, between Patrick and Peter Kavanagh, with Other Documents,* ed. P. Kavanagh (New York, 1969), 46.
32 A. Quinn, *Patrick Kavanagh: A Biography*, (Dublin: Gill & MacMillan, 2001), 125.
33 Kavanagh, *The Great Hunger*, 24.
34 M. Folch-Serra, "Place, Voice, Space: Mikhail Bakhtin's Dialogical Landscape," in *Environment and Planning D*, 8 (1990): 258.
35 Kavanagh, "Ploughman," in *Collected Poems*, 1.

36 Kavanagh, *The Great Hunger*, 29.

37 Kavanagh, *The Green Fool*, 11.

38 P. Duffy, "Patrick Kavanagh's Landscape," *Eire-Ireland: A Journal of Irish Studies,* 21, no. 3 (1986): 110.

39 Ibid., 110.

40 Ibid.

41 Eamon de Valera, Speech to the nation, broadcast on Radio Éireann, March 17, 1943.

42 Patrick Kavanagh, "Introducing 'The Great Hunger,'" in *November Haggard: Uncollected Prose and Verse of Patrick Kavanagh*, ed. Peter Kavanagh (New York: Hand Press, 1971), 15.

43 E. Soja, *Postmetropolis: Critical Studies of Cities and Regions* (Malden: Blackwell, 2000), xv, 13.

44 Ibid.

45 Kavanagh, "I Had a Future," in *Collected Poems*, 261–62.

46 Corrigan, "Qualitative GIS and Emergent Semantics," 82.

47 M.-P. Kwan, "From Oral Histories to Visual Narratives: Re-Presenting the Post-September 11 Experiences of the Muslim Women in the USA," *Social and Cultural Geography,* 9, no. 6 (2008): 653–69.

Chapter 5

Modeling and visualizing in GIS: The topological influences of Homer's *Odyssey* and Dante's *Inferno* on James Joyce's *Ulysses* (1922)

Joycean cartographies

A humanities GIS model makes it possible to visualize how the narrative and geographical influences of Homer's *Odysseus* and Dante's *Inferno*, together with *Thom's Official Directory of the United Kingdom of Great Britain and Ireland*, contributed to James Joyce's creation of a kaleidoscopic "verbal representation of Dublin" in *Ulysses*.[1] Joyce composed the epic novel while living in Trieste, Zürich, and Paris, as the First World War unleashed a cataclysm that swept through the nations of Europe during the second decade of the twentieth century. In self-imposed exile from Ireland, Joyce declared through the character of Stephen Dedalus that "history . . . is a nightmare from which I am trying to awake."[2] To reimagine the streetscapes of his native city, Joyce relied on memory as well as the family and friends with whom he corresponded. Mostly, however, he looked to a map of Dublin as his primary source: an insert map in the 1904 edition of *Thom's* directory, dated to the official statistical record of October 25, 1903. Perhaps inspired by the nightmarish history unfolding around him, Joyce declared that in *Ulysses* he aimed "to give a picture of Dublin so complete that if the city

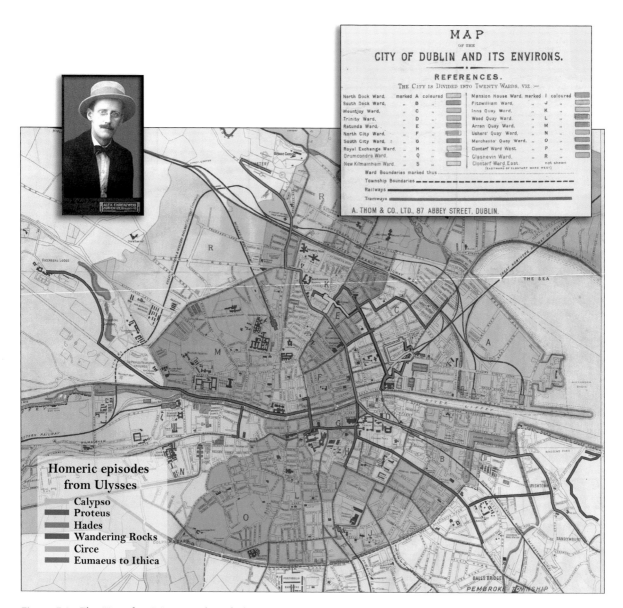

Figure 5.1   The "Joggfery" (geography) of Ulysses (1922). Created by the author from *Map of the City of Dublin and its Environs*, A. Thom & Co., Ltd., 87 Abbet Street, Dublin (1904), obtained from Trinity College Dublin Library; James Joyce, *Ulysses* (London: Penguin, 1992 [1922]); photo by Alex Ehrenzweig, 1915.

one day suddenly disappeared from the earth it could be reconstructed out of my book."[3] If one approaches the novel in a literal manner (and without prior knowledge of Irish history, literature, or culture), Joyce's depiction of the early twentieth-century city may appear as unintelligible; a rush of words in which space and time collide and clash in hopelessly fragmented, distorted, and incomprehensible ways. Joyce's structuring of *Ulysses'* narrative as a stream of consciousness (a literary device in which the interior monologues of a character melt into their environment and setting) may encourage this perception. However, Joyce, displaying all the precision of a master cartographer, mapped his novel by consulting *Thom's* directory and seeking corroboration from Dubliners:

> He was fanatic as well about verifying certain geographical minutiae, writing to his Aunt Josephine, for example, to inquire whether "an ordinary person" might climb over the area railing at 7 Eccles Street and safely lower himself down to gain entry through the lower level of the house.[4]

In such a manner, Joyce "remotely sensed" the layout and atmosphere of Dublin's streets, districts, pubs, churches, houses, and neighborhoods to create a mental map from which to plot the paths of his characters as their itineraries crisscrossed the city. Through techniques of literary cartography and montage, Joyce not only recreated a physical reality of Dublin (colored by his personality) but also the official, statistical city enshrined in *Thom's* directory, embalming by reference and allusion its street lists, tradesmen's catalogues, and census counts.[5]

Ulysses takes place on June 16, 1904, largely in the consciousness of Joyce's principal characters: the 22-year-old student Stephen Dedalus and 38-year-old advertising salesman Leopold Bloom. Both characters experience alienation in their daily lives. Dedalus mourns his mother (despite his refusal to pray at her deathbed), while Bloom, a Jew in a largely Catholic country, carries the dual burden of his son Rudy's recent death and the knowledge that his wife, Molly, is cuckolding him with a *bon vivant* named Blazes Boylan. Their separate peregrinations through the different districts and neighborhoods of Dublin during the day eventually converge in the book's phantasmagoric "Nighttown" section. Drawing on the skills of an "engineer" and a "scissors and paste man," Joyce used *Thom's* directory and map to erect Cartesian scaffolding over the city of Dublin. In this framework, and influenced by Classical Greek and Medieval Italian epic poetry, he stitched together the book's plotlines. Once he had sewn the fabric of his book together, Joyce dismantled the cartographic structure to reveal a kaleidoscopic textual tapestry. Humanities GIS techniques make it possible to creatively reconstruct Joyce's literary deconstruction of the *Thom's* map of Dublin to identify the sites that integrate the topologies and literary allusions from Homer and Dante into the novel.

Homer and Dante's topologies

Joyce based the wanderings of Dedalus and Bloom through Dublin on the voyages of the *Odyssey* (8 BC). Homer's epic poem describes Odysseus' adventures during the 10 years he traveled across the Aegean and Mediterranean seas in an effort to return home to Ithaca after the Trojan Wars.

Looking to Homer, Joyce divided *Ulysses* into three main sections—*The Telemachiad*, *The Odyssey*, and *The Nostos*—as narrative structures to plot the whereabouts of Bloom and Dedalus as they journey through Dublin's streets and districts.[6]

However, Joyce also looked to another literary work for a template to *Ulysses*: Alighieri Dante's epic poem *Divina Commedia* (ca. 1308–21). Reading the poet in college, he declared, "I love Dante almost as much as the Bible. He is my spiritual food"[7] and claimed that "Europe's epic is the *Divine Comedy*." A paperback edition of the poem—an undated 1904 reissue—provided Joyce with Dante's literary and historical sources through which he identified people, places, and events as well as allusions and geographical references.[8] The *Divine Comedy* presents a cosmography strikingly different than the one depicted by Homer in the *Odyssey*. Dante's poem tells the story of his descent with the ancient Roman poet Virgil into the three levels (nine circles) of suffering to the center of Hell, before ascending Mount Purgatory to Paradise. Influenced by the poem's first cantica, *Inferno*, Joyce plotted the paths that Bloom (as Virgil) and Dedalus (as Dante) forge across Dublin as they symbolically descend into Hell to the foot of Mount Purgatory on the doorstep of Bloom's house on Eccles Street in the early morning hours of June 17. There, the pair of voyagers glimpse the constellation of Paradise.[9]

In contrast to Homer's heroic geography, the *Inferno* imparts a medieval Catholic cosmology in which time and space assume eschatological and hierarchical functions of an ecclesiastical globe, cleaved by the pit of Hell. Topologically, Homer's narrative plots movement along a horizontal axis (on the plane of the Mediterranean and Aegean Seas), while Dante's narrative indicates a downward movement on a vertical axis (a descent into Hell). In this way, Joyce engaged the schemas and topologies of both narratives to chart Bloom's and Dedalus's journeys. Employing humanities GIS techniques, then, we can explore how Homer's and Dante's spatial and narrative influences converge in the fluid mosaic of Joyce's novel.

Modeling *Ulysses*

I incorporated *Thom's* directory map of Dublin into a humanities GIS model in figure 5.1 to facilitate three forms of interpretive visualization: hermeneutic (textual and topological), ergodic (mapping alternative narrative paths in a "cybertext"), and deformative (deliberate textual misreadings). This GIS model serves three objectives. The first seeks to create a topographical picture of Dublin using the cartographical source material that Joyce used to plot his novel. The second translates Joyce's "cut-and-paste" and other visual and literary methods into GIS form. The third explores and maps how Joyce narratively and topologically linked various Dublin locations to selected Homeric episodes with symbolic references to Dante's journey. I selected six of the eighteen Homeric episodes from *Ulysses* to map in this chapter. Using digital editing software, I took the six different sections of *Thom's* map of

Dublin that correspond with the six chosen Homeric episodes, digitized them into raster images, and imported them into ArcMap where I georectified the sections to create a GIS model for each episode.

1. In the model for each episode, I created a polyline shapefile tracing Bloom's and Dedalus's specific movement using GIS editing techniques based on Joyce's text and the schematic of the streetscapes displayed by the digitized sections of *Thom's* map displayed in the ArcMap window.

2. I then geocoded the sites and corresponding passages in the GIS model frames of each Homeric episode using point and polyline shapefiles to identify locations and to map the paths of Joyce's characters.

3. Last, I created three fields in the attribute tables of each episode's location point shapefile. The first field listed the ID of the episode, and the second and third fields listed textual linkages between *Ulysses* and *Inferno*. Though not shown in the following figures, this GIS function is useful for conducting spatial hermeneutic readings.

The attribute tables provide a means for scholars to make short annotations on the passages, locations, and character paths in the six chosen episodes. I enhanced this GIS interpretive approach by adding hyperlinks to a full-scale *Thom's* map using red circles to identify each episode's location in Dublin and text boxes that feature truncated and deformative *Wordle Cloud* readings of passages from each episode to visualize Joyce's stream-of-consciousness technique. The use of ArcMap and ArcCatalog in this manner does not so much serve to export a finished image or representation (although this is certainly an option) as it does to provide a tool to visualize, situate, and create topological hermeneutics as well as ergodic and deformative interpretive readings of a piece of literature.

The topologies of *Ulysses*

The GIS models of Homer's six episodes—(1) "Calypso," (2) "Proteus," (3) "Hades," (4) "Wandering Rocks," (5) "Circe," and (6) "Ithaca"—illustrate possible means to explore the influences of the *Odyssey* and *Inferno* on Joyce's novel. By no means definitive, these interpretive models provides the foundation for the ArcScene 3D visualization featured at the end of this chapter, illustrating how Homer's and Dante's schemas and topologies may have guided Joyce in plotting Bloom's and Dedalus's journey across Dublin on June 16, 1904. The following chapter sections discuss how GIS helped to both visualize and analyze the ways Joyce linked various locations in Dublin to the three main levels of Hell (upper, middle, and lower) and Purgatory in the *Inferno* to the six Homeric episodes of *Ulysses*.

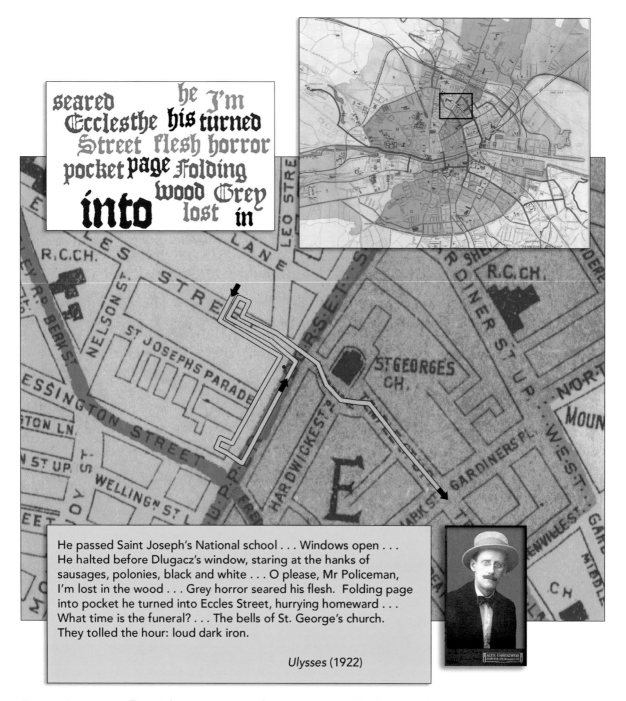

seared he I'm
Ecclesthe his turned
Street flesh horror
pocket page Folding
wood Grey
into lost in

He passed Saint Joseph's National school . . . Windows open . . .
He halted before Dlugacz's window, staring at the hanks of
sausages, polonies, black and white . . . O please, Mr Policeman,
I'm lost in the wood . . . Grey horror seared his flesh. Folding page
into pocket he turned into Eccles Street, hurrying homeward . . .
What time is the funeral? . . . The bells of St. George's church.
They tolled the hour: loud dark iron.

Ulysses (1922)

Figure 5.2 **Upper Hell, (1)** *Calypso*, **8 a.m. (7 Eccles Street).** Created by the author from *Map of the City of Dublin and its Environs*, A. Thom & Co., Ltd., 87 Abbet Street, Dublin (1904), obtained from Trinity College Dublin Library; James Joyce, *Ulysses* (London: Penguin, 1992 [1922]); photo by Alex Ehrenzweig, 1915.

Upper Hell

(1) *Calypso*, 8 a.m. (7 Eccles Street)

Although it is the fourth episode in the novel, "Calypso" is the first to introduce Leopold Bloom as he leaves his house on Eccles Street. Joyce writes that Bloom "ate with relish the inner organs of beasts and fowls,"[10] and the GIS model displays the path he takes walking to Dlugacz's butcher shop to buy a breakfast of "mutton kidneys which gave to his palate a fine tang of faintly scented urine."[11] Like Dante at the beginning of the *Inferno*, Bloom finds himself "lost in the wood."[12] Alienated as a Jewish cuckold in a Catholic country, his life corresponds with Dante's first *terza rima*:

> In the middle of the journey of our life
> I came to myself within a dark wood
> Where the straight way was lost.[13]

The geocoded word boxes and clouds in the GIS analysis of "Calypso" correspond with Bloom's path and his thoughts as he travels to and from 7 Eccles Street, to Dlugacz's butcher shop, and past St. George's Church[14] to begin his journey across Dublin (figure 5.2). The episode closes with Bloom's musings about Paddy Dignam's burial procession, which he will participate in during the "Hades" episode.

(2) *Proteus*, 11 a.m. (Sandymount Strand)

The GIS model of "Proteus" traces Stephen Dedalus's walk on the beach (shown in purple) as he follows the shoreline at Sandymount Strand and gazes out across the rippling mudflats at low tide (figure 5.3).[15] Joyce composed this episode as a stream-of-consciousness set piece. Through Dedalus's thoughts and impressions, Joyce tackles the philosophical question of how objects are perceived in time (*nacheinander*) and space (*nebeneinander*). Dedalus is haunted by the memory of his recently deceased mother, "a ghostwoman with ashes on her breath."[16] The burden of his mother's death on Dedalus's mind increases as he questions the nature of divinity and reality, wonders who his father really is, and grapples with financial solvency. Like Dante (and Bloom), he finds himself "*Where the straight way was lost*" and pondering the,

> Ineluctable modality of the visible: at least that if no more, thought through my eyes.
> Signatures of all things I am here to read, seaspawn and seawrack, the nearing tide, that
> rusty boot. Snotgreen, bluesilver, rust: coloured signs.[17]

Analysis of the GIS models of "Calypso" and "Proteus" confirm Clive Hart's and Leo Knuth's observation in *A Topographical Guide to James Joyce's Ulysses* (1975) that the "topography of Dublin is 'on the page' at least as much as are the meanings of the words . . . 'kidney', or 'ineluctable modality': it is part of the book's primary reference system, without which its full sense cannot be apprehended."[18] In "Proteus," Dedalus is glimpsed by Bloom, who in the 11 a.m. episode of "Hades" joins a funeral procession commencing outside Paddy Dignam's house in Sandymount.

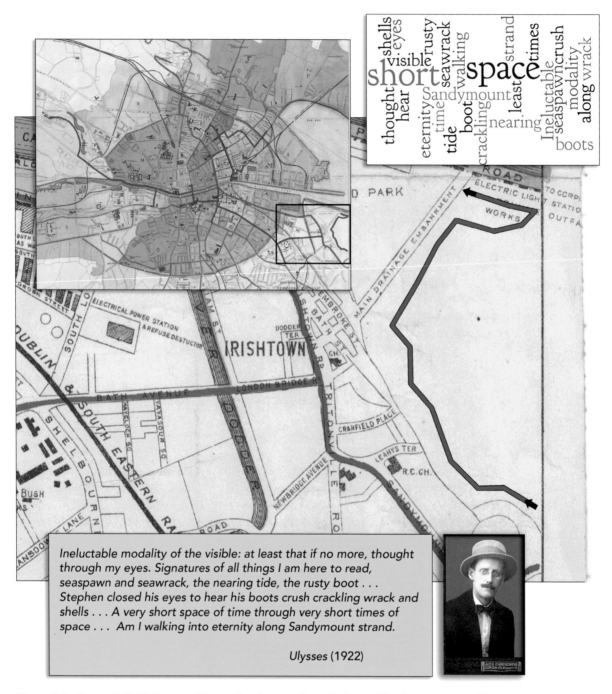

Ineluctable modality of the visible: at least that if no more, thought
through my eyes. Signatures of all things I am here to read,
seaspawn and seawrack, the nearing tide, the rusty boot . . .
Stephen closed his eyes to hear his boots crush crackling wrack and
shells . . . A very short space of time through very short times of
space . . . Am I walking into eternity along Sandymount strand.

Ulysses (1922)

Figure 5.3   **Upper Hell, (2)** *Proteus*, **11 a.m. (Sandymount Strand).** Created by the author from *Map*
of the City of Dublin and its Environs, A. Thom & Co., Ltd., 87 Abbet Street, Dublin (1904); obtained
from Trinity College Dublin Library; James Joyce, *Ulysses* (London: Penguin, 1992 [1922]). Photo by Alex
Ehrenzweig, 1915.

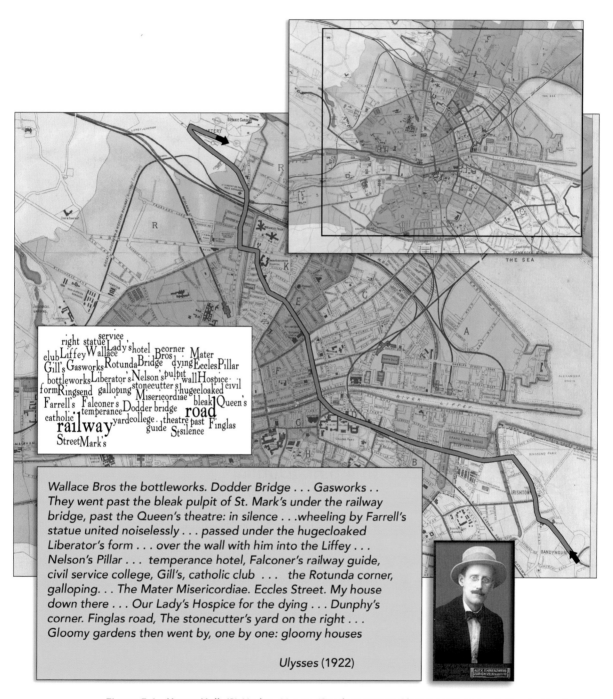

Wallace Bros the bottleworks. Dodder Bridge . . . Gasworks . .
They went past the bleak pulpit of St. Mark's under the railway
bridge, past the Queen's theatre: in silence . . .wheeling by Farrell's
statue united noiselessly . . . passed under the hugecloaked
Liberator's form . . . over the wall with him into the Liffey . . .
Nelson's Pillar . . . temperance hotel, Falconer's railway guide,
civil service college, Gill's, catholic club . . . the Rotunda corner,
galloping. . . The Mater Misericordiae. Eccles Street. My house
down there . . . Our Lady's Hospice for the dying . . . Dunphy's
corner. Finglas road, The stonecutter's yard on the right . . .
Gloomy gardens then went by, one by one: gloomy houses

Ulysses (1922)

Figure 5.4   **Upper Hell, (3) *Hades*, 11 a.m. (Sandymount to Glasnevin Cemetery).** Created by the
author from *Map of the City of Dublin and its Environs*, A. Thom & Co., Ltd., 87 Abbet Street, Dublin
(1904); obtained from Trinity College Dublin Library; James Joyce, *Ulysses* (London: Penguin, 1992
[1922]). Photo by Alex Ehrenzweig, 1915.

(3) *Hades,* 11 a.m. (Sandymount to Glasnevin Cemetery)

Martin Cunningham, Mr. Power, and Simon Dedalus, Stephen's father, join Bloom in the funeral carriage in Sandymount. The "Hades" episode presents an impressionary travelogue across the streetscapes of Edwardian Dublin. The GIS model charts the path of the funeral cortege (figure 5. 4) as it travels through the city from Sandymount in the southeast to Glasnevin Cemetery in the northwest, passing over the River Liffey.[19] The geocoded word box lists the sites that the cortege passes on its funeral journey.

- In the *Inferno*, Dante and Virgil cross the River Styx to the "Ring of the Wrathful" and enter into the City of Dis.
- *Ulysses* refers to crossing the River Liffey ("*over the wall with him into the Liffey*"[20]), which cuts through the center of Dublin.
- Bloom's passage through the City of Dis, represented in the following "Wandering Rocks" episode, occurs before he enters "Nighttown" and the "Center of Hell" with Stephen Dedalus in the "Circe" episode.

The GIS model visualizes the funeral cortege's dual function in facilitating an understanding of the contiguous nature and simultaneous play of the Homeric and Dantean schemas and topologies embedded in *Ulysses*.

Middle of Hell (City of Dis)

(4) *Wandering Rocks,* 2:55 to 4 p.m. (Dublin)

The GIS model of the "Wandering Rocks" episode illustrates the path of the Viceregal Calvacade from Phoenix Park north of the River Liffey toward Ballsbridge (figure 5.5) on the afternoon of 16 June 1904.[21] Although it acts as an analogy to the City of Dis, Frank Budgen observes, "this is peculiarly the episode of Dublin itself. Its houses, streets, spaces, tramways and waterways are shown us."[22] As a nationalist, Joyce once stated that he did not see "what good it does to fulminate against the English tyranny while Roman tyranny occupies the palace of the soul."[23] The episode opens and closes with ciphers of these institutions: the Jesuit Father John Conmee, who steps out of his church for a stroll and tram ride, and the Viceregal Calvacade carrying the Second Earl of Dudley as it clip-clops through the suburb of Ballsbridge to end it. Budgen recalls,

> Joyce wrote the Wandering Rocks with a map of Dublin before him on which were traced in red ink the paths of the Earl of Dudley and Father Conmee. He calculated to a minute the time necessary for his characters to cover a given distance of the city.[24]

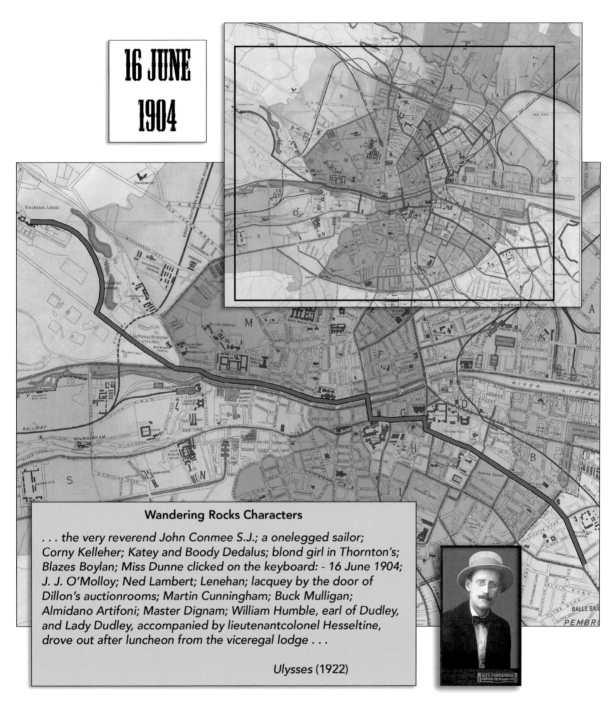

16 JUNE 1904

Wandering Rocks Characters

. . . the very reverend John Conmee S.J.; a onelegged sailor; Corny Kelleher; Katey and Boody Dedalus; blond girl in Thornton's; Blazes Boylan; Miss Dunne clicked on the keyboard: - 16 June 1904; J. J. O'Molloy; Ned Lambert; Lenehan; lacquey by the door of Dillon's auctionrooms; Martin Cunningham; Buck Mulligan; Almidano Artifoni; Master Dignam; William Humble, earl of Dudley, and Lady Dudley, accompanied by lieutenantcolonel Hesseltine, drove out after luncheon from the viceregal lodge . . .

Ulysses (1922)

Figure 5.5   Middle Hell (City of Dis), (4) *Wandering Rocks* **"Viceregal Calvacade," 2:55 to 4 p.m. (Dublin).** Created by the author from *Map of the City of Dublin and its Environs*, A. Thom & Co., Ltd., 87 Abbet Street, Dublin (1904); obtained from Trinity College Dublin Library; James Joyce, *Ulysses* (London: Penguin, 1992 [1922]). Photo by Alex Ehrenzweig, 1915.

The viewpoint of the episode changes from one sentence to another, forcing readers to be alert to follow the continuous variations of scale and angle, from close-up to a bird's-eye view. Budgen observes,

> A character is introduced to us at close-up range, and suddenly without warning, the movement of another character a mile distant is described. The scale suddenly changes. Bodies become small in relation to the vast space around them. The persons look like moving specks. It is a town seen from the top of a tower.[25]

This episode underlines the feasibility of employing GIS as a lens to explore the paths and patterns of Joyce's characters as they cross Dublin. Budgen also points out that Joyce was influenced by a game called *Labyrinth*, which he played nightly with his daughter, Lucia. Winning and losing the game "enabled him to catalogue six main errors of judgment into which one might fall in choosing a right, left or centre way out of the maze."[26]

A complex GIS model of "Wandering Rocks" could incorporate the spatial and logical schema from *Labyrinth*, the topologies of Homer and Dante's epic poems, and the street layout from *Thom's* map to analyze the paths of individual Dubliners as they spider web across the city in his book. In the *Inferno*, the City of Dis guards the entrance to Lower Hell, which is reserved for a panoply of sinners, including flatterers, simonists, barraters, soothsayers, hypocrites, thieves, robbers, evil counsellors, sowers of discord, and falsifiers. Apparently, Joyce's characters embrace and contest this spectrum of sins in varying scales and scopes. "Wandering Rocks" comprises a tragi-comedy of the human landscape of Edwardian, late colonial Dublin and hints simultaneously at the quotidian prospects of damnation and salvation. The episode stands in contrast to the sulphurous darkness of Nighttown in the "Circe" episode in which Bloom and Dedalus finally meet and descend to the Center of Hell.

Lower Hell

(5) *Circe*, 12 Midnight (Nighttown)

The GIS model of "Circe" (figure 5.6) visualizes the paths of Bloom and Dedalus as they finally cross in Bella Cohen's brothel after the older man follows the younger man from the railway station into the red-light district of "Nighttown" to save him from exploitation.[27] Dedalus, terrorized by the hallucination of his mother's rotting corpse rising from the grave, smashes the chandelier in Bella's parlor with his ashplant (walking stick) and flees, leaving Bloom to haggle over the damages. Declaring, "I need mountain air" (an allusion to Dante's desire to reach Mount Purgatory), Bloom rushes from the brothel out into the street, where Dedalus shouts in heated argument with two English soldiers, Carr and Compton: "[I]t is I who must kill the king and priest."[28] With a crowd watching, he receives a punch for his efforts. The police arrive, people disperse, and Bloom tends to the younger man. As he does so, he experiences a hallucination of his dead son, Rudy. With instances of textual free play, Joyce constructed "Circe" as a theater

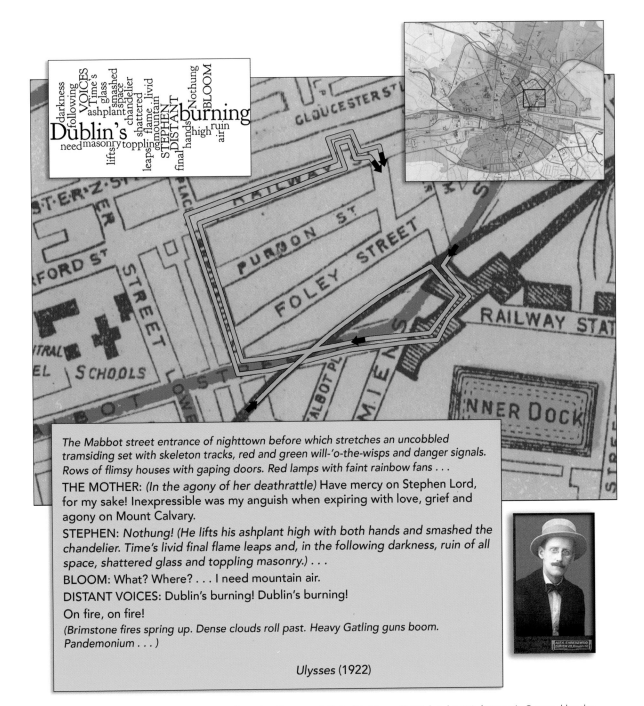

The Mabbot street entrance of nighttown before which stretches an uncobbled tramsiding set with skeleton tracks, red and green will-'o-the-wisps and danger signals. Rows of flimsy houses with gaping doors. Red lamps with faint rainbow fans . . .

THE MOTHER: *(In the agony of her deathrattle)* Have mercy on Stephen Lord, for my sake! Inexpressible was my anguish when expiring with love, grief and agony on Mount Calvary.

STEPHEN: *Nothung!* (He lifts his ashplant high with both hands and smashed the chandelier. Time's livid final flame leaps and, in the following darkness, ruin of all space, shattered glass and toppling masonry.) . . .

BLOOM: What? Where? . . . I need mountain air.

DISTANT VOICES: Dublin's burning! Dublin's burning!

On fire, on fire!

(Brimstone fires spring up. Dense clouds roll past. Heavy Gatling guns boom. Pandemonium . . .)

Ulysses (1922)

Figure 5.6 **Lower Hell to the Center of Hell, (5)** *Circe*, **12 Midnight (Nighttown).** Created by the author from *Map of the City of Dublin and its Environs*, A. Thom & Co., Ltd., 87 Abbet Street, Dublin (1904); obtained from Trinity College Dublin Library; James Joyce, *Ulysses* (London: Penguin, 1992 [1922]). Photo by Alex Ehrenzweig, 1915.

script complete with stage instructions and hallucinatory visions induced by epilepsy and absinthe.[29] The unconscious map of "Circe" is an ambiguous one because the episode functions as a microcosm and memory bank of *Ulysses*. Most of the characters, situations, and events in Joyce's novel reappear transposed and rearranged in the episode as distortions of their previous incarnations. Functioning like memory, the episode's connections, associative rather than linear, organize around images.[30] The episode includes messengers who serve as guides in the underworld of the unconscious, much like the damned in Dante's *Inferno*. The actual realm of the warren of streets that constitute the "Nighttown" district, "Circe" originates from a location in northeast Dublin (figure 5.6) where, Frank Budgen notes, Catholic tradition and police tolerance allowed a whole streetside of houses to function openly as brothels, with prostitutes sitting on doorsteps freely soliciting as if they were in Marseilles. Joyce informed Budgen that he employed "trivial and quadrivial words and local geographic allusions" in composing "Circe."[31] The episode's phantasmagorical ending invokes a Dantean hellscape that foreshadows the conflagration of the city ("*Dublin's burning! Dublin's burning!*") in the 1916 Rising.

Purgatory

(6) *Ithaca* (including Nostos's journey through "Eumaeus"), 2 to 4 a.m. (7 Eccles Street)

As the GIS model illustrates, Joyce's Dantean pair make their way out of the chaos and conflict of Nighttown and proceed on their journey through the "Eumaeus" episode, in which they stop to drink coffee in a cabman's shelter, to Bloom's house at 7 Eccles Street where "Ithaca" takes place. Correspondingly, in the *Inferno*, Dante and Virgil descend down to the ice-choked Lake of Cocytus in the ninth circle. Satan, encased in its waters to his waist, weeps bitter tears. They climb down his body and over his buttocks and escape through a passage that leads to the foot of Mount Purgatory, above which the constellations of Paradise rise in majesty:

> The Guide and I into that hidden road
> Now entered, to return to the bright world;
> And without care of having any rest
> We mounted up, he first and I the second,
> Till I beheld through a round aperture
> Some of the beauteous things that Heaven doth bear;
> Thence we came forth to re behold the stars.[32]

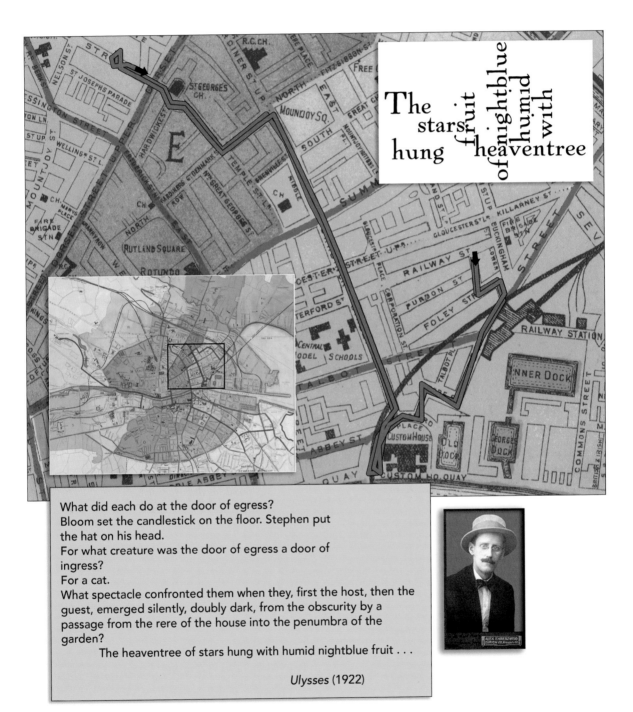

The stars hung heaventree of fruit nightblue humid with

What did each do at the door of egress?
Bloom set the candlestick on the floor. Stephen put the hat on his head.
For what creature was the door of egress a door of ingress?
For a cat.
What spectacle confronted them when they, first the host, then the guest, emerged silently, doubly dark, from the obscurity by a passage from the rere of the house into the penumbra of the garden?
The heaventree of stars hung with humid nightblue fruit . . .

Ulysses (1922)

Figure 5.7   Purgatory, (6) *Ithaca* **(including Nostos's journey through "Eumaeus"), 2 to 4 a.m. (7 Eccles Street).** Created by the author from *Map of the City of Dublin and its Environs*, A. Thom & Co., Ltd., 87 Abbet Street, Dublin (1904); obtained from Trinity College Dublin Library; James Joyce, *Ulysses* (London: Penguin, 1992 [1922]). Photo by Alex Ehrenzweig, 1915.

Bloom, having forgotten the latchkey, climbs down the railings to gain access to 7 Eccles Street. He then invites Dedalus inside, where he prepares two teacups of Epps soluble cocoa for himself and his guest.[33] The GIS model of their paths visualizes the "Ithaca" episode within a Euclidian universe, where parallel lines do not intersect. Tracking the pair from the cabmen's shelter to Eccles Street, Joyce shows that two human beings can come together only to the extent that they can approximate each other's courses but never can merge as one.[34] Bloom and Dedalus also represent two respective temperaments: the scientific and the artistic. However, both agree from their different perspectives on the "incertitude of the void" that characterizes human existence.[35] The word boxes in the GIS model contains literal and deformative readings of the passage from "Ithaca" that describes Dedalus leaving Bloom's house as daybreak approaches on June 17:

> What did each do at the door of egress?
> Bloom set the candlestick on the floor. Stephen put
> the hat on his head.
> For what creature was the door of egress a door of
> ingress?
> For a cat.
> What spectacle confronted them when they, first the host, then the guest, emerged silently,
> doubly dark, from the obscurity by a passage from the rere of the house into the penumbra
> of the garden?
> The heaventree of stars hung with humid nightblue fruit.[36]

A GIS-facilitated analysis of the relationship between *Thom's* map of Dublin, six Homeric episodes from *Ulysses*, and the Cantos of the *Inferno* make it possible to topologically link Joyce's "heaventree of stars" over Bloom's house on Eccles Street to the stars Dante witnesses as he emerges from purgatory in the *Divina Commedia*.[37] Here, GIS provides the user with the ability to illustrate a novel's narrative paths as well as collate, display, and analyze the different literary, cultural, and topological influences embedded in the text. In this manner, GIS is not being utilized simply as a map-producing machine but instead functions as a tool that collates, juxtaposes, and explores the relationships between a writer's text and various source materials to create hermeneutic, ergodic, and deformative interpretive visualizations of their work.

Visualizing a "new *Inferno* in full sail"

Frank Budgen recalled that "to see Joyce at work on the *Wandering Rocks*" section of his novel "was to see an engineer at work with compass and slide-rule, a surveyor with theodolite and measuring chain, or more Ulyssean perhaps, a ship's officer taking the sun, reading the log and calculating current drift and leeway."[38] The maritime analogy is apt;

after reading *Ulysses,* the poet Ezra Pound declared the novel "a magnificent new *Inferno* in full sail."[39] In addition to Joyce's literary innovations, Budgen observes that *Ulysses* contains traces of visual techniques employed in cubism, futurism, simultanism, and Dadaism. The "multiplicity of technical devices is proof that Joyce subscribed to no limiting aesthetic creed, and proof also that he was willing to use any available instrument that might serve his purpose."[40] One can only imagine what Joyce might have accomplished had he access to GIS to engage *Thom's* map of Dublin as he plotted his novel. It seems Joyce's artistic sensibilities would have inclined him to enjoy playing with the technology while drafting his work. Discussing the Cyclops episode with Budgen in 1919, Joyce asked him a question:

> "Does this episode strike you as being futuristic?"
> said Joyce.
> "Rather cubist than futurist," I said.
> "Every event is a many-sided object. You first state one view of it and then you draw it from another angle to another scale, and both aspects lie side by side in the same picture."[41]

To illustrate this Joycean perspective, ArcMap and ArcScene helped to create a 3D model that overlaid Homer's and Dante's schemas and topologies onto a digitized *Thom's* map of Dublin to visualize how Joyce conceptualized the voyages of Bloom and Dedalus.

The six Homeric episodes based on journeys taken over the Mediterranean and Aegean Seas in the *Odyssey* (figure 5.8) were drafted on a horizontal axis across the face of the city.

Figure 5.8   **"Homer's worldview."** From The Challenger Reports (summary), 1895.

Figure 5.9   **Sandro Botticelli's painting of Dante's levels of Hell from the *Inferno*.** By Sandro Boticelli, "Chart of Hell," circa 1480 to 1490.

Then, using Dante's levels of Hell as a guide (figure 5.9), the journey in *Ulysses* was simultaneously plotted, with a chronology running from eight o'clock on the morning of 16 June 1904 to four o'clock in the morning of 17 June, as a descent down a deep slope passing through the three levels of Hell (upper, middle, and lower) and ending at Bloom's house on Eccles Street in the "Ithaca" episode.

To visualize "*Ulysses* as a magnificent new *Inferno* in full sail" featured in figure 5.10, an ArcMap template hosted and displayed a digitized full-scale raster layer of *Thom's* map of Dublin as a base image.

- The six Homeric episode polyline patterns featured in the previous sections of this chapter overlaid this raster image, each highlighted in different colors by transparent polygon shapefiles.

- The shapefile creation and editing functions of ArcCatalog and ArcMap helped to draft three concentric-circle, polyline shapefiles with diminishing radii—to represent the upper, middle, and lower levels of Hell based on the topology of the levels of Hell in the *Inferno* (figure 5.9).

- The three concentric circles representing levels of Hell were edited in ArcMap so that the six Homeric episode patterns corresponded with their specific place in Dante's *Inferno* as referenced by Joyce in *Ulysses*.

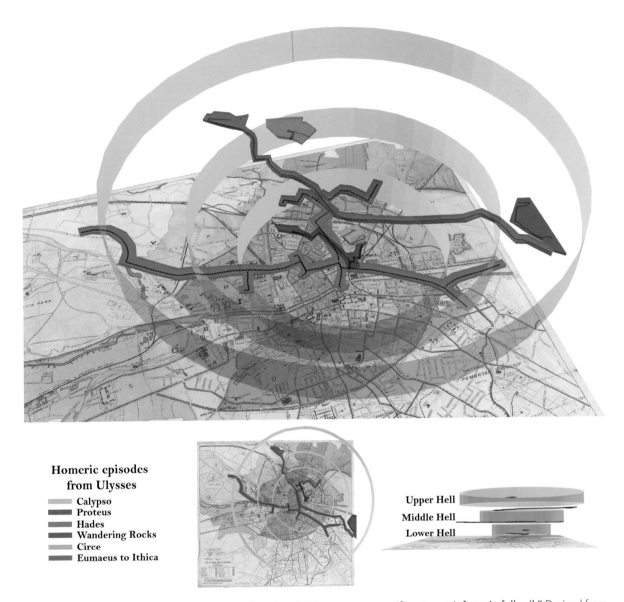

Homeric episodes from Ulysses

- Calypso
- Proteus
- Hades
- Wandering Rocks
- Circe
- Eumaeus to Ithica

Upper Hell
Middle Hell
Lower Hell

Figure 5.10   **ArcScene visualization of "*Ulysses* as a magnificent new *Inferno* in full sail."** Derived from *Map of the City of Dublin and its Environs*, A. Thom & Co., Ltd., 87 Abbet Street, Dublin (1904), from the Trinity College Dublin Library.

Once this was accomplished, I engaged ArcScene to import the rasterized map of Dublin and the shapefiles representing the Homeric episodes and the three circles of Hell.

1. I calculated the six Homeric episodes and circle of Hell shapefiles with different base heights and extruded them to create the 3D visualization of their intersection in *Ulysses*.

2. In the visualization, the Homeric episodes suspend in successive, descending, and chronological order in the upper, middle, and lower levels of Hell over their respective locations on the *Thom's* map of Dublin.

3. The Homeric episodes and three levels of Hell correlate with the time of day and location in which they took place on June 16 and 17 in Joyce's plotting of the novel.

The humanities GIS approaches in this chapter show that it is possible to visually engage Joyce's inter-textual dialogue between Homer and Dante to explore how the author transposed the theme of "history repeating itself with a difference" to Dublin. GIS made it possible to see how Joyce's appropriation of both the Greek and Medieval poets' schemas and topologies longitudinally shaped cross-sections of Bloom's and Dedalus's inner and outer journeys. It has been argued that the medium of Joyce's novel is wholly spatial, and using "Immersive and Experiential" GIS techniques, readers could find themselves strolling through the streets, districts, and squares of 1904 Dublin, walking in and out of pubs, shops, churches, and houses when and where they liked.[42] In conclusion, one can only imagine the fun Joyce may have had with GIS at his disposal. He may have found it a useful interactive tool for creating and rotating his own 3D model of Dublin, adding and deleting data and turning layers on and off, while plotting Bloom's and Dedalus' journey in *Ulysses* across space and time.

Sources

1   C. Hart and L. Knuth, *A Topographical Guide to James Joyce's* Ulysses (Colchester: A Wake Newsletter Press, 1975), 13.

2   J. Joyce, *Ulysses* (1922; repr. London: Penguin, 1992), 42.

3   F. Budgen, *James Joyce and the Making of "Ulysses"* (Oxford: Oxford University Press, 1972), 69.

4   M. Tymoczko, *The Irish Ulysses* (Berkley: The University of California Press, 1994), 154.

5   Hart and Knuth, *Topographical Guide to James Joyce's "Ulysses,"* 14.

6   Michael Seidel's, *Epic Geography: James Joyce's "Ulysses"* (1986) reconstructs the journey in the book to Bloom's house on Eccles Street in such a manner.

7   H. Helsinger, "Joyce and Dante," "*ELH*" 35, no. 4 (December 1968), 591–605.

8   R. Ellmann, *James Joyce* (New York, 1965); H. Helsinger, "Joyce and Dante," *ELH* 35, no. 4 (December 1968), 591–605; M. T. Reynolds "Joyce's Edition of Dante," *James Joyce Quarterly*, 15, no. 4 (Summer 1978), 380–84.

9   C. Slade, "The Dantean Journey Through Dublin," *Modern Language Studies,* 6 , no. 1 (1976): 12.

10 Joyce, *Ulysses*, 65.

11 Ibid.

12 Ibid.

13 A. Dante, *The Divine Comedy*, vol. I, *Hell* (1949; repr. London: Penguin, 1974), 1–3.

14 Joyce, *Ulysses*, 65, 67, 70, 71, 73, 85.

15 Ibid., 45.

16 Ibid., 46.

17 Ibid., 45.

18 Hart and Knuth, *Topographical Guide to James Joyce's "Ulysses,"* 18.

19 Joyce, *Ulysses,* 108, 109, 112, 113, 114, 116, 119, 121, 122, 123, 125, 126.

20 Ibid., 126.

21 Joyce, *Ulysses,* 280, 288, 289, 291, 294, 296, 297, 301, 304, 318, 319, 321, 323, 324.

22 Budgen, *James Joyce and the Making of "Ulysses,"* 124–25.

23 James Joyce, "Ireland, Island of Saints and Sages," lecture, Università Popolare, Trieste (April 27, 1907), in *James Joyce: Occasional, Critical and Political Writing,* ed. K. Barry (Oxford: Oxford University Press, 2002), 125.

24 Budgen, *James Joyce and the Making of "Ulysses,"* 124–25.

25 Ibid., 126.

26 Ibid., 125.

27 Joyce, *Ulysses,* 561, 683, 684, 694.

28 Ibid., 688.

29 J. F. Lowe, "Circean Aerodynamics," *Texas Studies in Literature and Language,* 51, no. 4; "News of Ulysses: Readings and Re-Readings," (Winter 2009): 476–93.

30 Robert D. Newman, "The Left-Handed Path of 'Circe,'" *James Joyce Quarterly,* 23, no. 2 (Winter 1986): 223–27.

31 Newman, "The Left-Handed Path of 'Circe'"; F. Budgen, *James Joyce and the Making of "Ulysses,"* 125.

32 Dante, *The Divine Comedy, 133–139.*

33 Joyce, *Ulysses,* 788.

34 Avrom Fleishman, "Science in Ithaca," *Wisconsin Studies in Contemporary Literature,* 8, no. 3 (Summer 1967): 377–91.

35 Joyce, *Ulysses,* 798, 818.

36 Ibid., 818–19.

37 Slade, "The Dantean Journey," 18.

38 Budgen, *James Joyce and the Making of "Ulysses,"* 123.

39 Ezra Pound, letter to Homer Pound dated April 20, 1921, in *Pound/Joyce: The Letters of Ezra Pound to James Joyce, with Pound's Essay on Joyce,* ed. F. Read, (New York: New Directions, 1967), 189.

40 Budgen, *James Joyce and the Making of "Ulysses,"* 198.

41 Ibid., 156–57.

42 M. K. Booker, "From the Sublime to the Ridiculous: Dante's Beatrice and Joyce's Bella," *James Joyce Quarterly,* 29, no. 2 (Winter 1992): 357–68; A. Hauser, *The Social History of Art,* vol. 4 (London, 1972).

Chapter 6

Psychogeographical GIS: Creating a "kaleidoscope equipped with consciousness," Flann O'Brien's *At Swim-Two-Birds* (1939)

The novel as urban GIS

To borrow a phrase from Charles Baudelaire, Flann O'Brien's portrait of Dublin in *At Swim-Two-Birds* (1939) appears as though it was generated by a "kaleidoscope equipped with consciousness."[1] O'Brien's comical portrait of a 1930s university student consists of inter-lapping polychronic spaces and narrative lines converging to create a *flâneur's* literary impression of the city. The preceding chapters explain how scholars can apply GIS to the study of history and literature. In contrast, this chapter demonstrates how we can use a writer's text to conceptualize a humanities GIS model. Consequently, O'Brien's novel is considered as a blueprint from which a 3D, psychogeographical GIS model of urban postmodernity can be created. In turn, by adopting the counter-mapping strategies of the Situationist Internationalists, and focusing Giambattista Vico's cyclical view of history and Mikhail Bakhtin's chronotope through the lens of a GIS, *At Swim-Two-Birds* is transformed into a guide to explore how a writer's perceptions of a city can transpose layers of social interaction, myth, memory, fantasy, and desire upon its official cartography.[2] Critics have observed that O'Brien's "novel within a novel" reflects the concentric enfolding of postmodern urban events, suggesting that a cyclical collision of history and geography in Dublin, with its fiercely independent villages and suburbs, may have served as the template for the novel's multiple narrative lines and spaces.[3] It has also been argued that O'Brien's novel is a mind game that he plays on his readers, and its publication on the eve of the Second World War coincided with an increasing interest in game theory from economists, philosophers, computer scientists, and military strategists.[4] Given these contexts, one way to approach O'Brien's novel with GIS is to adopt William Cartwright's cartographical metaphors of "The Gameplayer" and "The Storyteller." These involve the creation of map-building games and digital stories in a GIS, in which navigation skills through simulated landscapes and virtual worlds can offer clues in solving puzzles about a particular location's sense of place.[5] In addition, Mei-Po Kwan and Guoxiang Ding have employed 3D GIS techniques to visualize the spaces of social interaction and time past and present to create geo-narratives about a storyteller's place.[6] In the following sections, a brief introduction to *At Swim-Two-Birds* precedes the chapter's discussion on how Situationist International, Viconian, Bakhtinian, as well as gaming and storytelling mapping strategies and techniques were operationalized in GIS to model the critical, spatial, and temporal dimensions of O'Brien's novel.

Spatializing *At Swim-Two-Birds*

O'Brien's portrait of 1930s Dublin takes place largely in the mindscape of a Dubliner and is rooted in the Irish storytelling tradition of the Medieval Irish romance of *Buile Suibhne* (The Rage of Sweeney) and the Celtic-Age epic *Táin Bó Cúailnge* (The Cattle Raid of Cooley). The day-to-day spaces of 1930s Dublin in *At Swim-Two-Birds* are occupied by O'Brien's student narrator who lives in his uncle's house and is writing a novel. The student spends the majority of his time in his bedroom "observing the street-scene arranged below" and lying on his bed smoking cigarettes. Enrolled in University College Dublin at Earlsfort Court Terrace, the student drinks in a public house adjacent to Stephen's Green named Grogan's, with his two true friends, Brinsley and Kelly. He also attends the cinema, bets on horses, and takes late evening walks across Dublin without purpose, save for the aim of "embracing virgins"[7] to kill time in the drab and depressing city in which he lives. The quotidian pace of everyday life, depicted through the eyes of the student, is marked by a persistent regularity:

> Nature of daily regime or curriculum: Nine thirty a.m. rise, wash, shave and proceed to breakfast; this on the insistence of my uncle, who accustomed to regard himself as the sun of his household, recalling all things to wakefulness on his own rising.
>
> 10.30. Return to bedroom.
>
> 12.00. Go, weather permitting, to College, there conducting light conversation on diverse topics with friends, or with acquaintances of a casual character.
>
> 2.00 p.m. Go home for lunch.
>
> 3.00. Return to bedroom. Engage in spare-time literary activity, or read.
>
> 6.00. Have tea in company with my uncle, attending in a perfunctory manner to the replies required by his talk.

7.00. Return to bedroom and rest in darkness.

8.00. Continue resting or meet acquaintances in open thoroughfares or places of public resort.

11.00. Return to bedroom.

Minutiae: No. Of cigarettes smoked, average 8.3; glasses of stout or other comparable intoxicant, av. 1.2; times to stool, av. 2.65; hours of study, av-1.4; spare-time or recreative pursuits, 6.63 circulating.[8]

The temporal schema of this narrative strand echoes Henri Lefebvre's early studies of the everyday, analyzing the false consciousness and alienation emerging from pop culture and the growing consumerism of the 1930s.[9] In contrast, Dublin's surreal and metaphysical layers in *At Swim-Two-Birds* are transposed upon its cityscape in the novel that the student is composing. Its storyline draws on *The Cattle Raid of Cooley* and *The Rage of Sweeney* and concerns Dermot Trellis, a publican, and erstwhile novelist, whose characters inhabit the city streetscapes and districts of Lower Leeson Street, Sandymount, Irishtown, Ringsend, and the Palace Cinema on Pearse Street. Trellis' fictional characters include the cowboys Shorty Andrews and Slug Willard, a demonic figure named the Pooka MacPhellimey, and the working-class figures of Furriskey, Shanahan, and Lamont, who banter with the legendary hero of old Celtic Age Ireland, Finn Mac Cool, about the crazed mythological King Sweeney. The confluence of 1930s everyday life in Dublin and the metaphysical in O'Brien's novel illuminates Lefebvre's observation that unconscious desires and passions lie dormant beneath the surface of what appears real in the city and operate in the realm of the *surreal*.[10] The novel concludes with Trellis's cast of characters (who have mutinied and sentenced their author to death) being destroyed when his maid burns his manuscript to kindle a fire; the surreal then collapses to the gloom of 1930s Dublin, and we find the student, though barely having attended his classes, graduating from university to the surprise of his uncle. By resetting heroes from the ancient epics in a modern, urban setting, along with figures from Hollywood pop culture, and employing the use of parody, humor, and irreverence, O'Brien captured the dynamic bustle, flux, and personality of Dublin in the second decade of southern Ireland's independence from Britain.

Psychogeographical mapping with GIS

My strategy to map *At Swim-Two-Birds* with GIS drew on principles established by the Situationist International (SI) founded by Guy Debord in 1957. The SI was formed to counter hegemonic planning and architectural practices its members perceived were blighting the cityscape of Paris following the Second World War. The SI created its own vocabulary to investigate and articulate urban phenomena, coining the terms *dérive* (acts of urban drifting) and *psychogeography* (the emotional effects of a geographical environment) to describe a couple of their mapping tactics. Acts of urban drifting according to Debord reveal that cities have a psychogeographical relief and are roiling with constant emotional currents, fixed points, and vortices.[11] Mapping this type of phenomena consists of hijacking and modifying versions of official maps, urban plans, or blueprints to highlight emotional digressions from the status quo.[12] Drafting a psychogeographical map can be considered a game that undermines the supposed objectivity of Cartesian perspectivalism to examine the complex topographical relationships between mental, emotional, and social space.[13] Using GIS to spatially parse the text of O'Brien's novel by hijacking a rasterized copy of the 1931 Ordnance Survey (OS) Map of Dublin, I engaged in what the SI call an act of *détournement*. This involves integrating artistic and other works to produce a superior construction of a milieu, in this case, a postmodern critical literary perspective of Dublin in the 1930s.

Intrigued by how open-source digital platforms such as *Second Life* allowed students to study and enact Shakespearean plays in a virtual 3D recreation of the Globe Theater, I loaded my laptop computer with Esri software to create a 3D, psychogeographical GIS model of *At Swim-Two-Birds* to conduct and physically perform literary analysis in the virtualized streets of Dublin.[14] Imagining myself as an avatar for O'Brien's student narrator, I employed the metaphors of the game player and storyteller and took to the city's streets, various districts, sites, and public houses to conduct an urban survey based on SI principles and practices. To create this GIS virtual-map game, I imported a digitized copy of an Ordnance Survey Map of Dublin from 1931 into ArcMap as a raster image, which I carried into the field on my laptop. This allowed me to experience a convergence between the textual and virtual Dublins and the actual ambiences of the city. Using *At Swim-Two-Birds* as a guide and expending a modest amount of shoe leather, I followed, surveyed, and geocoded the locations and paths connecting the sites in O'Brien's novel as they uncoiled across the city.

To create the GIS model visualization featured in figure 6.2, I began my psychogeographical plotting of O'Brien's novel at the site where the colonial "eye" of Nelson's Pillar (a British imperial relic destroyed in the 1960s) on O'Connell Street once looked out over the cityscape of Dublin in the 1930s (figure 6.1, no. 1). In the novel, the student leaves his uncle's house in a moment of pique and walks heatedly to the pillar. Acting as the student's avatar, I commenced mapping this spot and followed his literary footsteps into the postcolonial city of O'Brien's imagination:

- From the pillar, I drifted south across O'Connell Bridge over the River Liffey and made my way to St Stephen's Green. There, I located Grogan's Pub (figure 6.1, no. 3) on the southwestern corner of the green.
- O'Brien's student narrator frequented this pub, as did his friends, so I established this location, adjacent to the student's university at Earlsfort Terrace (figure 6.1, no. 2) as the site to anchor their walks in *At Swim-Two-Birds* across Dublin's streetscapes.

Using GIS, I geocoded the various locations described in O'Brien's novel visited by the student narrator and his friends. By collating and linking the possible routes that they could have followed, given the street layout of Dublin, a spatial narrative of the character's movements to certain locations and their links to O'Brien's surreal retelling of *The Cattle Raid of Cooley* and *The Rage of Sweeney* were identified:

- For instance, in one section of O'Brien's novel (figure 6.1, no. 5), the student narrator meets his friend Kelly at the corner of Kildare and Nassau streets.

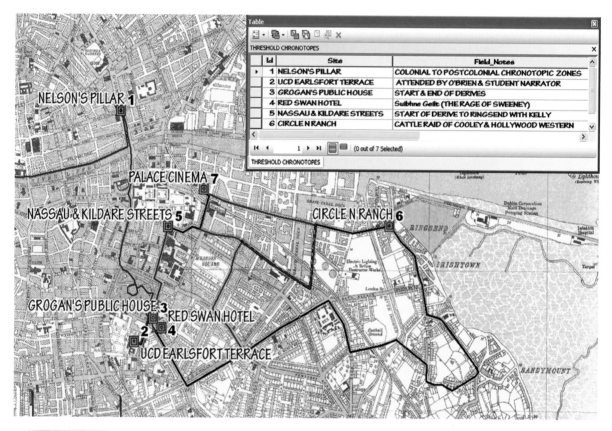

Id	Site	Field_Notes
1	NELSON'S PILLAR	COLONIAL TO POSTCOLONIAL CHRONOTOPIC ZONES
2	UCD EARLSFORT TERRACE	ATTENDED BY O'BRIEN & STUDENT NARRATOR
3	GROGAN'S PUBLIC HOUSE	START & END OF DERIVES
4	RED SWAN HOTEL	*Suibhne Geilt* (THE RAGE OF SWEENEY)
5	NASSAU & KILDARE STREETS	START OF DERIVE TO RINGSEND WITH KELLY
6	CIRCLE N RANCH	CATTLE RAID OF COOLEY & HOLLYWOOD WESTERN

Mental/Physical Dérives through Dublin in *At Swim Two Birds* (1939)

Figure 6.1 **Mental/physical dérives through Dublin in *At Swim-Two-Birds* (1939), with an accompanying photo of Flann O'Brien celebrating Bloomsday from a carriage window (1954).** Created by the author from the Saorstát Eireann Ordnance Survey (OS) Dublin & Environs 1: 20000 Sh 265b, published in 1934, from the Trinity College Dublin Library. Photo courtesy of National Library of Ireland, from the Wiltshire Photographic Collection, published/created June, 1954.

- They then walk to the site of the Circle "N" Ranch (figure 6.1, no. 6) located in the south Dublin neighborhoods of Ringsend, Irishtown, and Sandymount.

Employing the vernacular of Hollywood and pulp-fiction Westerns, O'Brien re-enacts the Celtic-Age saga *The Cattle Raid of Cooley* by having 1930s corner boys in Ringsend throw stones at an electric streetcar in a game of cowboys and Indians. The student and his friend Kelly then return to Grogan's Pub and walk by the fictional Red Swan Hotel (figure 6.1, no. 4) on Lower Leeson Street. At this site, O'Brien retells the story of the medieval epic *The Rage of Sweeney.*

According to *Thom's Official Directory of Dublin* for the years 1930–39, an establishment listed as the Eastwood Hotel occupied the address 91–92 Leeson Street, just half a block southeast of Grogan's. In O'Brien's novel, the student's character, the erstwhile novelist Dermot Trellis, is the proprietor of the Red Swan, where he lodges the characters of his own novel: "There is a cowboy in Room 13 and Mr Mc Cool, a hero of old Ireland, is on the floor above. The cellar is full of leprechauns."[15] In retelling *The Rage of Sweeney,* O'Brien reimagines Finn Mac Cool, a "hero of old Ireland," relating "the account of the madness of King Sweeney" to the working-class characters of Furriskey, Lamont, and Shanahan in the Red Swan Hotel.[16] These characters later hold a jury trial for Trellis at the Palace Cinema on Pearse Street (figure 6.1, no. 7). Like the SI's practice of psychogeography, O'Brien's novel draws on the shifting emotional perceptions of his characters as they move through Dublin's cityscape to depict the moods and different atmospheres of the city's various districts and neighborhoods. For instance, figure 6.1 illustrates the paths of dérive connecting psychogeographical sites where the student and his friends spy a gang of corner boys whose horseplay in the streets is the curse of the Ringsend district; the dangerous houses to be demolished in Irishtown and Sandymount; the Palace Cinema, packed with people, each with a cold-watching face, its air heavy and laden with sullen banks of tobacco smoke; Nassau Street, a district frequented by the prostitute class; the quiet of the ruined garden at Croppies Acre; and finally, Grogan's licensed premises, where, sitting on stools imbibing glasses of stout, the student and his friends experience a sense of physical and mental well-being.

Vico-Bakhtin timespaces

Spatially, a composite pattern emerges when the various walks taken by the student and his friends are plotted in GIS on the 1931 OS map of Dublin. On the map, the urban drift of the student and his friend in O'Brien's novel appears as a clockwise spiral that leads from Grogan's Pub to the corner of Dawson and Nassau Streets, to the ranch, to the hotel, and then back to the pub again and finally to the cinema. I then exported the map featured in figure 6.1 into ArcScene, in which I engaged its 3D Analyst function to transpose GIS visualizations of Bakhtin's chronotopic prism and Giambattista Vico's cyclical arcs of history over this pattern. This step allowed the collated narrative movements in *At Swim-Two-Birds* to be analyzed not only as a composite walk across the horizontal plane of Dublin in the 1930s but also as a vertical time travel journey through Viconian cycles of history.

Vico's *Scienza Nuova* (1725) conceptualizes a nonlinear perspective of history, which posits that societies ascend through three distinct ages before collapsing in a *ricorso* and commencing a new cycle. Vico sees this socio-temporal phenomenon as spiral, not repetitive, in the sense that social and cultural iterations occur whenever a new cycle begins. In *De Italorum Sapientia* (1710), Vico critiqued aspects of René Descartes methodology as it related to "civic life" by stating "to introduce geometrical method into practical life is 'like trying to go mad with the rules of reason,'

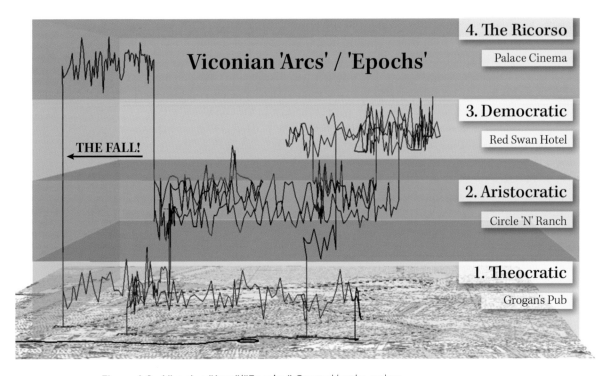

Figure 6.2 **Viconian "Arcs"/"Epochs."** Created by the author.

attempting to proceed by a straight line among the tortuosities of life, as though human affairs were not ruled by capriciousness, temerity, opportunity, and chance."[17] Figure 6.2 offers a GIS visualization of O'Brien's characters' paths (black polylines) through Vico's four arcs of history, featured as four different colored rectangular polygon layers extruded in successive height over the rasterized 1931 OS map of Dublin. The layers represent Vico's perspective on history, which envisions successive "arcs" or "epochs" that societies pass through in a cyclical fashion:

1. The first arc layer represents the *Theocratic* Age, when gods speak to mankind through oracles.
2. The second arc layer represents the *Aristocratic* Age, when the ruling classes come to dominate their societies without the need for divine sanction.
3. The third arc layer represents the *Democratic* Age, inaugurated by the "rational" and popularly elected governing assemblies.
4. The fourth and final arc layer represents the *Ricorso*—the age of demagoguery, social anomie, anarchy, and chaos—marked by *"la barbarie della riflessione* (the barbarism of reflection)" and eventual cultural decline and collapse.

The "fall" in the fourth arc signals a return to the *Theocratic* Age and heralds the birth of a new cycle of history for a given society. The clockwise spiral pattern was also extruded so that the student and his friend's walks to the various sites depicted in the 1931 OS map would correspond to a specific Viconian cycle of history. On the four successive Viconian arc layers, heroes from Celtic mythology and figures from American pulp-fiction Westerns and cinema simultaneously blend with the social geographies of petite bourgeois and working-class Dubliners to create the novel's ever-changing mélange of place. The multiple narrative lines in O'Brien's novel act in confluence with Bakhtin's chronotopes to illustrate that *At Swim-Two-Birds* depicts the completion of a Viconian cycle of history in the space of the city. In the GIS visualization (figure 6.3), chronotopes serve as nodes that connect the various paths that O'Brien's characters create as they move across the space of Dublin and gateways that facilitate their "time travel" up and down Viconian layers. As discussed in chapter 4, Bakhtin's chronotopes tie narrative strands together and thicken time by wrapping flesh upon its abstracted skeleton. Consequently, literary and actual space interacts with the forces of plot and history to act as x-rays of the culture from which they spring.[18] Because various sites in *At Swim-Two-Birds'* act as nodes and gateways, the novel's dominant chronotope is perceived as the:

> threshold and related chronotopes—those of the staircase, the front hall and corridor, as well as the chronotopes of the street and square [. . .] places where crisis events occur, the falls, resurrections, renewals, epiphanies.[19]

The pub, cinema, hotel, street, and districts identified in the GIS visualization (figure 6.4) intersect chronotopically with the four Viconian layers and are metaphorical gateways linking these sites in Dublin with different epochs. In the GIS visualization, chronotopes symbolize the "psychogeographical pivotal points" at play in the city in the novel. Through these time-spaces O'Brien's student narrator oscillates up and down, passing through the "threshold" of each Viconian arc, making past and present in Bakhtin's words "essentially instantaneous."[20]

In the 3D, psychogeographical GIS model, it is possible to trace the urban drift of the novel's student narrator and his friend across the horizontal plane of Dublin, in tandem with their vertical time travel jags up and down the Viconian layers, as they complete a surreal cycle of history within the spaces of the city and the novel:

1. On the first layer (the "Theocratic Age"), O'Brien's biographical experience informs the novel's 1930s, day-to-day storyline that follows the student who is living with his uncle, writing a novel, attending University College Dublin at Earlsfort Terrace, peregrinating through various neighborhoods, and drinking in Grogan's public house on the corner of Leeson Street and Stephen's Green with his friends. Chronotopically, Nelson's Pillar acts as a gateway that ushers the reader from colonial to postcolonial space and to Grogan's Pub, which serves as the novel's anchor.

2. On the second layer (the "Aristocratic Age"), through O'Brien's retelling of the Celtic-era epic, *The Cattle Raid of Cooley*, we discover the Circle "N" Ranch and the cowboys Shorty and Slug trying to stop cattle from being rustled by "hooligans and corner boys" in the Irishtown, Ringsend, and Sandymount Districts of South Dublin. This occurs in the student's imagination as he passes through the district.

3. On the third layer (the "Democratic Age"), viewers find in the Red Swan Inn of Lower Leeson Street (Eastwood Hotel) its proprietor, Dermot Trellis, the publican-cum-author,

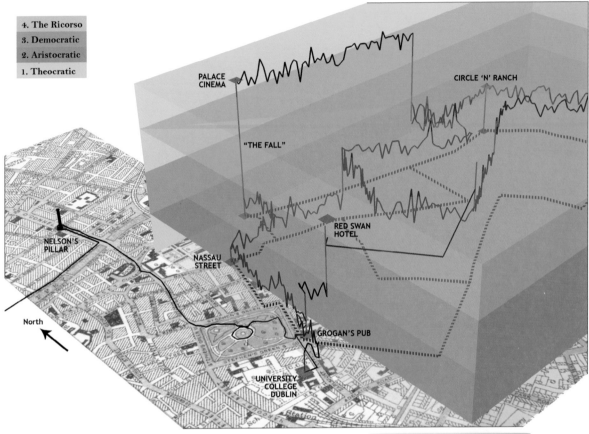

4. The Ricorso
3. Democratic
2. Aristocratic
1. Theocratic

Chronotope	Legend of sorts
Nelson's Pillar	Colonial to Postcolonial
University College Dublin	Biographical
Grogan's Pub	Start and end of many dérives
Red Swan Hotel	The Rage of Sweeney
Nassau Street	The Prostitute Class
Circle 'N Ranch (Ringsend)	Cattle Raid of Cooley
Palace Cinema	Trellis' Trial and the Fall!!!!

Figure 6.3 **Viconian ages and chronotopic portals.** Created by the author from Saorstát Eireann
Ordnance Survey (OS) Dublin & Environs 1: 20000 Sh 265b, published in 1934. Obtained from Trinity
College Dublin Library.

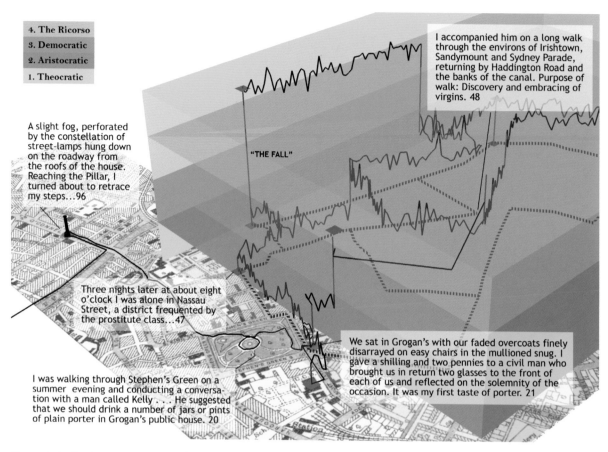

Figure 6.4   Student narrator's psychogeographies in *At Swim-Two-Birds* (1939). Created by the author from Saorstát Eireann Ordnance Survey (OS) Dublin & Environs 1: 20000 Sh 265b published in 1934. Obtained from Trinity College Dublin Library.

the other characters of the student's novel, including King Sweeney, living in congress and playing poker. Again, this is a site the student passes and embellishes with his imagination.

4.   On the fourth layer (the "Age of Ricorso"), we find Trellis's fictional characters, led by the rebel Pooka MacPhellimey, setting up an ad-hoc courtroom in the Palace Cinema on Pearse Street, where they judge and sentence him to death. Trellis is saved when his servant accidently burns his manuscript, which precipitates the-novel-within-a-novel's fiery denouement and the "Viconian Fall" back to the "Theocratic Age" of 1930s Dublin.

The novel's other chronotopic portals also facilitate the student and his friends' walks across the city and allow them to oscillate across the mental-spatial-temporal boundaries of Dublin's four Viconian ages operating in *At Swim-Two-Birds*. The 3D, psychogeographical GIS model focuses Bakhtin's chronotopic lens so that:

> Time and space merge [...] into an inseparable unity [and] the plot (sum of depicted events) and the characters [...] are like those creative forces that formulated and humanised this landscape, they made it a speaking vestige of the movement of history (historical time).[21]

Counter-cartographical GIS

Gunnar Olsson notes that "in French the word *représentation* means both 'representation' and 'performance.'"[22] The 3D, psychogeographical GIS model of Dublin, based on the textual blueprint of *At Swim-Two-Birds*, fulfills both definitions of the word. The surrealism of O'Brien's urban portrait bears a similarity to Alfred Döblin's *Berlin Alexanderplatz* (1929), which used techniques pioneered by James Joyce and John Dos Passos to paint kaleidoscopic pictures of Berlin.[23] O'Brien's work reimagines Irish identity in an urban setting by borrowing—with "astonishing eclecticism"—from "cowboy novels, proletarian balladry, racing tipsters, encyclopaedias and modernist literature."[24] This recalls Roland Barthes' contention that a literary text constitutes "a multi-dimensional space in which a variety of writings, none of them original," blend and clash. Furthermore, the postmodern narrative structures of *At Swim-Two-Birds* provide a means to conceptualize how users can model linear and cyclical schemas of time and space simultaneously in GIS. By linking the novel's various paths, networks, patterns, and interactions, the GIS model exploits the physicality and the landscape of the novel's pages, capturing the concurrencies and conjunctures of place, which often escapes the linear language of much literary criticism.[25] It also illuminates, through the endless repetition of daily paths in an urban world, the ever same reproduced as the ever new.[26] This subversive form of cognitive GIS modeling departs from positivistic-scientific rigor to provide the colors of a creative and subjective type of cartographical art.[27] The extrusion of Vico and Bakhtin's theoretical schemas in three-dimensions over the 1931 OS map of Dublin illustrate Debord and the Situationist International's contention that ordinary street plans provide fixed spatial fields on which we can identify the entrance and exits to the psychogeographical pivot points of a city.[28] The GIS model facilitates ergodic and deformative readings of O'Brien's novel, and by linking these various pivot points to the OS map of Dublin, illuminates Olsson's observation that:

> Every map is a palimpsest, a many layered imagination of another place and another time. Translation is nevertheless the name of the game, for to translate is to make new boots out of old ones, to convey an idea from one art form to another.[29]

In conclusion, the 3D visualizations of 1930s Dublin created by the psychogeographical GIS model of *At Swim-Two-Birds* underscores Lefebvre's observation in *La révolucion urbaine* that "each place and moment has no existence except within the ensemble, the contrasts and oppositions that link them to the other places and moments they are distinguished from."[30] This echoes *Hercules Furens*, the epigraph from O'Brien's novel: "All things naturally draw apart and give place to one another."

Sources

1   Benjamin, *Illuminations*, 171.

2   D. L. Parsons, *Streetwalking the Metropolis: Women, the City and Modernity* (Oxford: Oxford University Press, 2003).

3   J. Hassett, "Flann O'Brien and the Idea of the City," in *The Irish Writer and the City*, ed. M. Harmon (Gerrard's Cross: Colin Smythe, 1984); D. Kiberd, *Irish Classics* (London: Granta, 2001).

4   K. Bohman-Kalaja, "The Truth Is an Odd Number: At Swim-Two-Birds," in *"Is It about a Bicycle?" Flann O'Brien in the Twenty-First Century*, ed. J. Baines (Dublin: Four Courts Press, 2011).

5   W. Cartwright, "Extending the Map Metaphor Using Web Delivered Multimedia," *International Journal of Geographical Information Science,* 13 (1999): 335–53.

6   M.-P. Kwan and G. Ding, "Geo-Narrative: Extending Geographic Information Systems for Narrative Analysis in Qualitative and Mixed-Method Research," *The Professional Geographer,* 60, no. 4: 443–65.

7   F. O'Brien, *At Swim-Two-Birds* (London: Penguin, 1939), 12, 38.

8   Ibid., 148–49.

9   Henri Lefebvre, *Critique de la vie quotidienne*, vol. 1, *Introduction* (Paris: Grasset, 1947).

10 A. Merrifield, *Henri Lefebvre: A Critical Introduction* (London and New York: Routledge, 2006).

11 G. Debord, "Introduction to a Critique of Urban Geography," in *Situationist International Anthology*, ed. Ken Knabb (Berkeley: Bureau of Public Secrets, 1981), 50.

12 Ibid., 50–54.

13 P. Mitchell, *Cartographic Strategies of Postmodernity: The Figure of the Map in Contemporary Theory and Fiction* (London and New York: Routledge, 2008), 120.

14 P. Cohen, "Humanities 2.0, Giving Literature Virtual Life," *The New York Times*, March 21, 2011, http://www.nytimes.com/2011/03/22/books/digital-humanities-boots-up-on-some-campuses.html?ref=humanities20.

15 O'Brien, *At Swim-Two-Birds*, 35.

16 Ibid., 36.

17 "Giambattista Vico," in *New World Encyclopedia* [online], accessed March 17, 2013, http://www.newworldencyclopedia.org/entry/Giambattista_Vico.

18 Bakhtin, *The Dialogic Imagination*, 81, 250, 425–26.

19 Ibid., 248.

20 Ibid.

21 Bakhtin, "Speech Genres and Other Late Essays," 49.

22 G. Olsson, *Abysmal* (Chicago: The University of Chicago Press, 2007), 135.

23 D. Macey, *The Penguin Dictionary of Critical Theory* (London: Penguin, 2001), 119.

24 Ibid., 502.

25 A. Pred, *Lost Words and Lost Worlds: Modernity and the Language of Everyday Life in Late-Nineteenth Century Stockholm* (Cambridge: Cambridge University Press, 1990), xv–xvi; D. Gregory, *Geographical Imaginations* (Cambridge, MA: Blackwell, 1994).

26 Gregory, *Geographical Imaginations.*
27 Mitchell, "Cartographic Strategies of Postmodernity," 119.
28 Debord, "Introduction to a Critique of Urban Geography," 50.
29 Olsson, *Abysmal,* 135.
30 Lefebvre, *La révolucion urbaine,* 53–54.

Chapter 7

Geovisualizing Beckett

Samuel Beckett's GIStimeline

Irish Nobel laureate writer Samuel Beckett (1902–89) rendered affective, comedic, and existential depictions of Dublin, London, and Saint-Lô during his European travels in the 1930s and 1940s that illustrate a shift in his literary perspective from a latent Cartesian verisimilitude to a more phenomenological and fragmented, existential impression of place. GIStimeline visualizations of this period of *Wanderjahre* provide a geographical and historical context to his intellectual and aesthetic influences, which shaped what would later become known as the Beckettian literary landscape. Visualizing Beckett's biographical trek along a GIStimeline contextualizes his early pieces of poetry and fiction in the period during which:

> there was a formative relationship between literary innovation and the cross-cultural status of many modernist and avant-garde artists, those who during the first half of the century came to London, Paris or Berlin from "colonized or capitalized regions [within Europe] . . . linguistic borderlands . . . [or] as exiles . . . from rejecting or rejected political regimes."[1]

The website, the *Digital Literary Atlas of Ireland, 1922–1949*, hosts integrated Esri GIS and open-source, geospatial technology–produced interfaces, as well as EXHIBIT Timeline functions made available by the Semantic Interoperability of Metadata and Information in unlike Environments (SIMILE) project, developed by the Massachusetts Institute of Technology's (MIT) Computer Science and Artificial Intelligence Laboratory (CSAIL) and Library. In this open source–enabled GIS platform, period images from the streetscapes of Dublin, London, Paris, and Saint-Lô were hyperlinked to the Samuel Beckett page of the digital atlas, so interactive visualizations of his travels could be produced to illustrate the shift in Beckett's literary perspective from the Cartesian to the phenomenal.[2]

The GIStimeline also provides a cinematic mobility contiguous with the period in which Beckett was undertaking his formative travel. It is also conversant with the internationalism and aesthetic modes of the Italian Futurists and German Expressionists whose images presaged the kinetic ocular medium of the World Wide Web and its dynamic geospatiality.[3] Readers may wish to

Figure 7.1   *The Digital Samuel Beckett.* Created by the author from Saorstát Eireann Ordnance Survey (OS) Dublin & Environs 1: 20000 Sh 265b, published in 1934. Obtained from Trinity College Dublin Library. Digital Atlas image, courtesy of Trinity College Dublin. Google and the Google logo are registered trademarks of Google Inc., used with permission.

access the digital atlas and navigate their own ergodic and deformative visual and textual journeys to explore Beckett's biography and its relation to his writing style and its connection to place.

Geovisual narratology

It has been noted that "geographic visualization works by providing graphical ideation to render a place, a phenomenon or a process visible, enabling the most powerful human information-processing abilities—those of spatial cognition associated with the eye-brain vision system—to be directly brought to bear."[4] The proliferation of digital tablet reading devices provides the perfect interface to synthesize visual and textual narratives in a historically contingent and contextual geospatial frame to facilitate the study of writers, period, and place. According to Andrew Thacker, social and literary spaces both operate in relation to historical coordinates. He further emphasizes that not only do literary texts represent social spaces but also, in turn, social spaces influence the very shape of narrative forms. The task of a critical literary geography is to trace the ways in which historical and social spaces intrude on the internal construction of literature.[5]

Creating interactive geographic visualizations with GIS constitutes a post-structuralist form of literary authorship and analysis. David Staley observes that far from reinforcing the classic Aristotelian linear narrative, GIS facilitates a multidimensional narrative space of images and words. Constructing or reading a geographic visualization in GIS is like composing or parsing an ergodic story because it requires labor from both the procedural author and the active viewer to both chart and navigate its potential narrative paths.[6]

In this sense, GIS provides a means to "reconnect the representational spaces on literary texts not only to material spaces they depict, but also reverse the moment" and consider the influence of actual places on shaping a writer's perspective and work.[7] Once again, GIS makes it possible to chart the trajectories of a story's characters and perspectives and to plot both the resulting cultural centers and the multiple social, environmental, and imaginative dimensions of space against the historical poetics of place that emerge in the narrative lines of the writer's work. This is important because it activates the process of mapping relations between variables, during which unexpected connections can produce insights that elude conventional linear and analog forms of biography and literary analysis. By integrating geospatial interpretations of text into such a platform, the procedural author of a GIStimeline visualization serves simultaneously as both a kind of "narrative architect" and a viewer of a spatial story.[8] This in turn facilitates ergodic and deformative approaches that allow the author-viewer to juxtapose different scales of time, space, and text to explore literary and historical perception. Traditionally, people depended on paper maps and dusty globes to represent geographic relationships, but the digital revolution has introduced interactive platforms and tools that we can use to create geospatial and geovisual narratives as well as "game" them, ergodically and deformatively. Stuart Aitken and James Craine have argued that geovisualization practitioners must refuse to take vision at face value and should insist on problematizing, theorizing, critiquing, and historicalizing the visual process.[9]

The GIStimeline platform illustrated in figure 7.2 facilitates a historicalized, interactive space, comprising a proprietary open-source digital architecture, to visualize Beckett's biography.

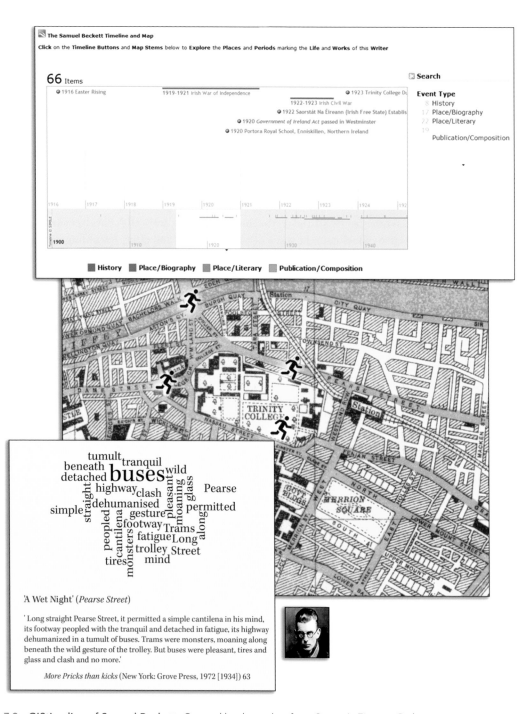

The Samuel Beckett Timeline and Map

Click on the Timeline Buttons and Map Stems below to Explore the Places and Periods marking the Life and Works of this Writer

66 Items

Search

1916 Easter Rising   1919-1921 Irish War of Independence   1923 Trinity College Du

1922-1923 Irish Civil War

1922 Saorstát Na Éireann (Irish Free State) Establis

1920 *Government of Ireland Act* passed in Westminster

1920 Portora Royal School, Enniskillen, Northern Ireland

Event Type
8 History
17 Place/Biography
22 Place/Literary
19
   Publication/Composition

■ History   ■ Place/Biography   ■ Place/Literary   ■ Publication/Composition

'A Wet Night' (*Pearse Street*)

' Long straight Pearse Street, it permitted a simple cantilena in his mind, its footway peopled with the tranquil and detached in fatigue, its highway dehumanized in a tumult of buses. Trams were monsters, moaning along beneath the wild gesture of the trolley. But buses were pleasant, tires and glass and clash and no more.'

More Pricks than kicks (New York: Grove Press, 1972 [1934]) 63

Figure 7.2 **GIStimeline of Samuel Beckett.** Created by the author from Saorstát Eireann Ordnance Survey (OS) Dublin & Environs 1: 20000 Sh 265b, published in 1934. Obtained from Trinity College Dublin Library.

Viewers can switch between his biographical timeline, whose windows contain snippets and deformative readings of his work, and GIStimeline framings of the place at which the two coincide (in this, an early piece, *A Wet Night,* which concerns a student's perception as he walks down Pearse Street, adjacent to Trinity College Dublin).

Figure 7.3 features an ArcGIS Map window that hosts a digitized historical map (upper left) hyperlinked to an interactive Samuel Beckett timeline hosted by MIT's EXHIBIT application, (middle) which visualizes code from a digitally published Google Docs spreadsheet (lower right) containing metadata of Beckett's biography. This type of open-source GIStimeline platform integrates the cartographic spaces of a map with textual and visual narratives tailored to a viewer's expertise.[10] Its interactive features reiterate the fact that geographic visualization is a cultural process that creates knowledge rather than merely reveals it.[11] The GIStimeline approach promotes a digital aesthetic that enables the user to write spatial narratives and create visual/ textual impressions and associations on the fly, in contrast to more scientifically focused GIS approaches. Beckett himself informs us that:

> An impression is for the writer what an experiment is for the scientist—with this difference, that in the case of the scientist the action of the intelligence precedes the event and in the case of the writer follows it.[12]

The interactive and dynamic nature of the GIStimeline function provides a cinematic mobility. Users can identify and map Beckett's biographical sites and render them into Keyhole Markup Language (KML) files and import them into ArcMap. This allows the user, through the use of ArcGIS animation features, to create further scales of knowledge and insight. The visualizations in this chapter used ArcGIS Online to show how it integrates with GIStimeline mappings. The release of ArcGIS Online offers immense possibilities for developers to integrate multiple geospatial platforms. As Aitken and Craine observe, by using GIS in such a manner,

> anyone can visually construct—or deconstruct—a spatial reality, and the result is a profound experiential and epistemological shift undergone by an increasingly digital culture. Referentially and legitimation are finished with construction of space: there is no longer an unproblematic and empirically verifiable "real" to refer *to*. What is left is esthetic and affective.[13]

Beckett, reflecting on the early twentieth-century developments of avant-garde visual art, once asked, "Must literature alone be forever left behind on worn out paths abandoned long ago by music and painting?"[14] The same can be asked about GIS approaches to spatial narrative. Indeed, scholars have long recognized that Beckett's work "is strikingly similar to a non-Euclidian geometry."[15]

By integrating the open source–enabled GIStimeline with ArcGIS Online, a digital scholar can engage in the kind of visual bricolage described by Claude Lévi-Strauss as a technique of improvisational collage. In this way, GIS analysts can crop, edit, geocode, and hyperlink Euclidian framings, photographs, textual extracts, and "Wordle" snippets to create a non-Euclidian, geovisual-temporal narrative that contextualizes Beckett's biographical lifepath with

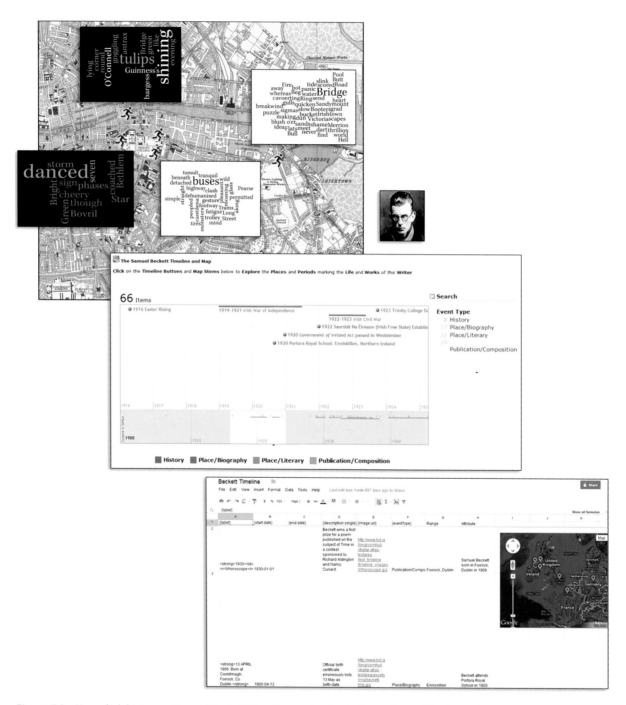

Figure 7.3 Hyperlink between Samuel Beckett timeline and web platform. Created by the author from Saorstát Eireann Ordnance Survey (OS) Dublin & Environs 1: 20000 Sh 265b, published in 1934. Obtained from Trinity College Dublin Library. Digital Atlas image courtesy of Trinity College Dublin. Google and the Google logo are registered trademarks of Google Inc., used with permission.

his literary works. The resulting mash-up analysis of Beckett's early works and experience of Dublin, London, Paris, and Saint-Lô in Normandy can—depending on choices made by the user—help users uncover meaningful patterns and reinterpretations of the relations between his literary landscapes, lifepath, and period.

The following sections illustrate one ergodically mapped and deformatively structured geovisual narration of Beckett's biography.[16] This is just one of many possible interpretations enabled by the GIStimeline.

Dublin-Paris, 1916–30

As a son of the Anglo-Irish *fin de siècle* bourgeoisie, Beckett witnessed at the age of 14 the Easter Rising in Ireland, a conflict symptomatic of wider historical and cultural cataclysms sweeping across Europe during the early decades of the twentieth century. Indeed, for most of his lifetime, the sight of Dublin ablaze during the week of Easter 1916, as he and his father watched from the hills above the city, remained deeply impressed in his mind.[17] Consequently, images of cities in ruins and denuded existential landscapes would come to serve as backdrops for his later works of fiction and drama. Beckett's portrayals convey Nigel Thrifts's idea that cities may be perceived as roiling maelstroms of affect. Particular affects are continually on the boil, and these affects continually manifest themselves in events that can take place either at a grand scale or as simply as a part of everyday life.[18] The operator of an open source–enabled GIStimeline can employ video-gaming techniques by merging interactive visualizations to create innumerable and dynamic story lines that zoom in and out at different scales. With practice, the user can make the pixels of a GIS computer screen change hue to form digital images of dynamic landscapes—aerial photographs if you will—that scale from the planetary sphere down to just above the desert floor.[19] Using such techniques accords with an observation Beckett made in 1934: "it is the act and not the object of perception that matters."[20] A GIS bricolage visualization of Beckett's timeline illustrates that,

> Perception is no longer anchored by the vanishing point in representation. It drifts in a landscape with no horizon. The affective interface is about a certain tension between physical and digital matter and space.[21]

In previous studies of writers, geographers have recognized that literature is the product of perception, or put more simply, is perception itself, and forms the basis for a new awareness, a new consciousness.[22] Furthermore, embodied practices of perception and writing produce a body of literature that can form the basis for a new and cleansed perception.[23] The evolution of Beckett's perceptual practices as a writer established him as a leading figure in twentieth-century literature for just such reasons. His works:

> do not want, by means of the conventional "verisimilitudes" of prose narrative, to impose order and structure upon a world in which, as they experience it, there actually is no order

or structure. This is why so often they deliberately refuse to look at anything other than the surfaces of the world.[24]

Figure 7.4 illustrates that much of the 1920s served as an incubation period for Beckett as a writer. Here, along the GIStimeline of his biography, we see the social and political turbulence borne from the Irish War of Independence (1919–21) and Irish Civil War (1921–22). He matriculated in 1923 to Trinity College Dublin, where he read French and Italian. While there, he immersed himself deeply in Dante Alighieri's *Divine Comedy*, and toward the decade's end, Beckett's significant early pieces begin to appear.

In 1928, Beckett won a two-year appointment to a fellowship as *lectuer d'anglais* at the École Normale Supérieure in France. During this period, he honed his interest in the history of spatial thought. Jean Beaufret, an expert on the phenomenologist Martin Heidegger, encouraged the young

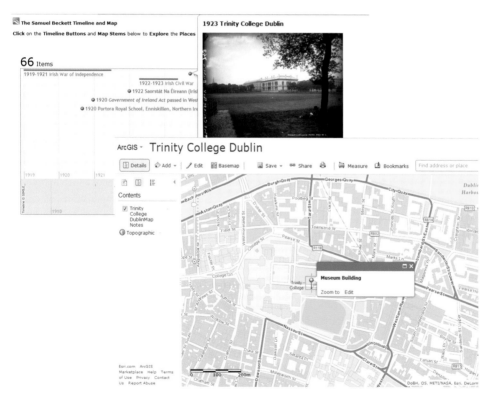

Figure 7.4 Trinity College Dublin. Photo by Robert French, chief photographer of William Lawrence Photographic Studios of Dublin. Digital Atlas image courtesy of Trinity College of Dublin. Data from ArcGIS Online, courtesy Esri, HERE, DeLorme, TomTom, Intermap, increment P Corp., GEBCO, USGS, FAO, NPS, NRCAN, GeoBase, IGN, Kadaster NL, Ordnance Survey, Esri Japan, METI, Esri China (Hong Kong), swisstopo, MapmyIndia, © OpenStreetMap contributors, and the GIS User Community.

Beckett to study René Descartes (1596–1650), the French mathematical philosopher whose Cartesian perspective symbolizes the dividing point between medieval and modern spatial sensibilities.[25] Descartes' *Cogito, Ergo Sum* influenced Beckett's perception that all existence remained in the head and all external contact was illusory. Beckett also pursued Gottfried Leibniz (1646–1716) for his relative notion of social space and Arnold Geulincx (1624–69) for occasionalism, which holds that the mind and body, though separate, synchronize like a film and its sound track.[26] Beckett's studies of these thinkers influenced his writing style; by extension, his interest in their different perspectives offers lessons about the various ways we can write space with GIS.

Between 1928 and 1929, Paris witnessed a proliferation of small, private presses and alternative journals that benefited Beckett's early literary efforts. The GIStimeline (figure 7.5) shows the location of Shakespeare & Co., a bookshop owned by Sylvia Beach, which Beckett frequented. The interface contains a photograph of James Joyce standing in its doorway. Beach published Joyce's *Ulysses* in 1922. In 1929, she published Beckett's essay, *Dante... Bruno. Vico..Joyce.*

During this time, Beckett also wrote for the journal *transition*, whose manifesto decried "THE HEGEMONY OF THE BANAL WORD, MONOTONOUS SYNTAX, STATIC PSYCHOLOGY, DESCRIPTIVE NATURALISM."[27] He frequented the Parisian coffee shop and bar milieu of writers and artists of the interwar "lost generation" and immersed himself in the philosophical ideologies of Nietzsche, Freud, and phenomenologists, such as Heidegger.[28] This marked his writing with a dizzying mix of the political Left and Right and schools of art, such as futurism, Dadaism, surrealism, residual cubism, and early abstract expressionism.[29] In 1930, Beckett won first prize for a poem on the subject of time, titled *Whoroscope*, which centered on the life of Descartes. The award led to a commission for a study entitled *Proust*, in which Beckett would dissect the modern concept of time. This work would, as a consequence, influence Beckett's increasingly fragmented representation of space in his subsequent works.

Figure 7.5 **Shakespeare & Co., Paris, with its location highlighted in a corresponding map.** Digital Atlas image courtesy of Trinity College Dublin. Data from ArcGIS Online, courtesy Esri, DigitalGlobe, GeoEye, i-cubed, USDA, USGS, AEX, Getmapping, Aerogrid, IGN, IGP, swisstopo, and the GIS User Community. Photographer unknown.

Beckett's bottled climates

Beckett returned to Dublin to assume a lectureship in French at Trinity College in 1930. However, Beckett concluded at the end of the Michaelmas term that his teaching position was a grotesque comedy.[30] Living in a city-center room at the college, he felt socially alienated from the Dublin *literati* and the cultural nationalists infusing the zeitgeist of the Free State. He found comfort only in his long walks, during which, he wrote, "the mind has a most pleasant and melancholy limpness, is a carrefour of memories, memories of childhood mostly, *moulin a larmes* [mill of tears]."[31] The walks were often spurred on in an effort to confront writer's block:

> As he endured long days with nothing to show but an occasional phrase or snatch of verse, his frustration grew, until he found himself spending most of his time walking (his remedy for coping with and possibly inviting the muse) or sitting in one pub after another until he abandoned any attempt at schedule or routine in disgust. Later, he would use these long frustrating walks—from one end of Dublin to the other, through the Wicklow Hills, along country lanes and past deserted railway stations—in his writing, in descriptions of the countryside or of his thoughts while pacing.[32]

During these walks, Beckett acted as a phenomenological surveyor, creating his own idiosyncratic GIS that produced discursive maps consisting of snatches of prose and poetic fragments, distinct and affective renderings of place. In a 1930 essay titled *Proust,* Beckett observed that the great French writer accepted "the sacred ruler and compass of literary geometry" and hinted that writing, like mapping, consisted perhaps in the "the imposition of our own familiar soul on the terrifying soul of our surroundings."[33] He also commented on the visual dimension of James Joyce's prose:

> It is to be looked and listened to. His writing is not *about* something; *it is that something itself* [...] When sense is sleep, the words go to sleep [...] when the sense is dancing, the words dance.[34]

Proust and Joyce influenced the shape of the visual and affective dimensions of his writing. Beckett described *More Pricks Than Kicks* (1934) as a collection of "bottled climates."[35] The stories convey vivid impressions of Dublin, as seen through the eyes of the marginalized Belacqua Shuah. Beckett based the character on a Florentine lute maker from Canto IX of Dante's *Purgatorio,* whose sloth and indolence in life have condemned him to spend time on the Mountain of Purgatory. His Belacqua, a slothful student of Dante at Trinity College, maintains that "the reality of the individual . . . is an incoherent reality and must be expressed incoherently."[36] Although Beckett conceived the character Belacqua during the period in which he was rootless both geographically and professionally, his native city and the Joycean influence left a strong imprint on the collection:[37]

> The Dublin background of *More Pricks Than Kicks* is carefully documented after the manner of *Ulysses:* the street names, the Liffey, Trinity College and the statue of Thomas Moore, combined to present the busy city landscape against which Belacqua is drawn.[38]

Beckett's impressionistic depiction of his native city illustrates modernity's implication of a specifically urban phenomenal world, one that was markedly quicker, more chaotic, fragmented, and disorienting than it was in previous phases of human culture.[39] The GIStimeline visualizes Dublin sites in both *More Pricks Than Kicks* and *Echoes Bones*, a collection of poems, to convey and incorporate contemporary perspectives on Beckett's work.

Beckett's physical and mental discomfort can be felt in his 1931 piece *Enueg II*. The poem's alienated and anxious speaker stands alone on a bridge in the angst-ridden heart of the city:

> lying on O'Connell Bridge
> goggling at the tulips of the evening
> the green tulips
> shining around the corner like an anthrax
> shining on Guinness's barges.[40]

The GIStimeline visualization of the bridge in figure 7.6 enables tablet-equipped readers to literally visit the site, in addition to other sites that inspired Beckett. They can, in the words of Annie Proulx, stand metaphorically in both the unwritten and the written landscapes and enter the territory of the page at the same time they created it in their minds—a profound engagement with place through both real 3D landscapes and also landscapes described and imagined.[41]

The visualization also provides a means to create immersive and experiential GIS interfaces that allow scholars to virtually navigate a writer's environment while remotely studying the places depicted in his literature through digital space.

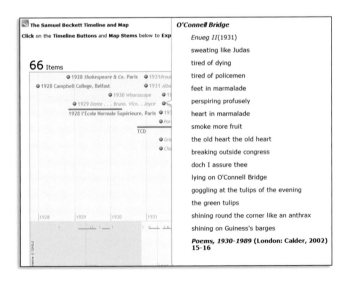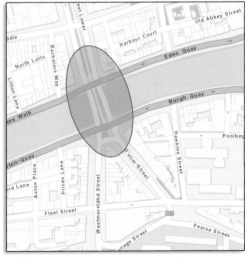

Figure 7.6   **O'Connell Bridge, Dublin.** Digital Atlas image courtesy of Trinity College Dublin. Data from ArcGIS Online, courtesy Esri, HERE, DeLorme, USGS, Intermap, increment P Corp., NRCAN, Esri Japan, METI, Esri China (Hong Kong), Esri (Thailand), TomTom, MapmyIndia, © OpenStreetMap contributors, and the GIS User Community.

Returning to the short story *A Wet Night*, the GIStimeline visualization offers a contemporary bird's-eye view of the bus-clogged entrance to "Long straight Pearse Street," where civil offices, places of commerce, and traffic imprint themselves on Belacqua's consciousness as he makes his way along:

> its footway peopled with the tranquil and detached in fatigue, its highway dehumanised
> in a tumult of buses. Trams were monsters, moaning along beneath the wild gesture of the
> trolley. But buses were pleasant, tires and glass and clash and no more.[42]

Beckett often began Belacqua's peregrinations through the streets of Dublin near the Thomas Moore Statue and the "hot bowels of McLouglin's"[43] pub that adjoin the front

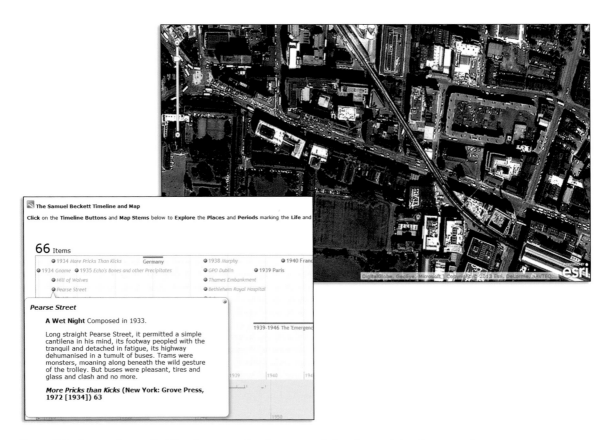

Figure 7.7   **Pearse Street, Dublin, shown running from upper left to lower right.** Courtesy of Trinity College Dublin. Data from ArcGIS Online, courtesy of Esri: World_Imagery (MapServer), Esri, DigitalGlobe, GeoEye, i-cubed, USDA, USGS, AEX, Getmapping, Aerogrid, IGN, IGP, swisstopo, and the GIS User Community.

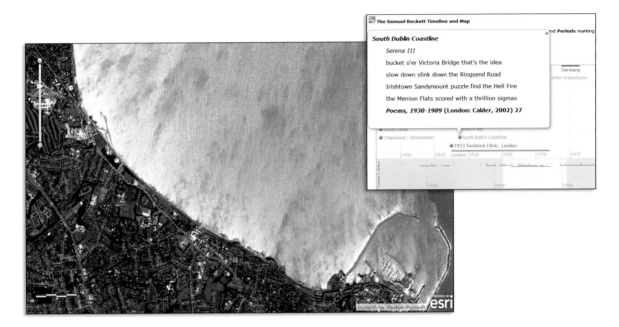

Figure 7.8 South Dublin coastline and corresponding GIStimeline for Samuel Beckett in the 1920s and 1930s. Courtesy of Trinity College Dublin. Data from ArcGIS Online, courtesy of Esri, DigitalGlobe, GeoEye, i-cubed, USDA, USGS, AEX, Getmapping, Aerogrid, IGN, IGP, swisstopo, and the GIS User Community.

gates of Trinity College, whose motto—*Perpetuis futuris temporibus duraturum* (It will last into endless future times)[44]—underscored Beckett's sense of time's repetitious nature. *Serena III*, also set in Dublin, describes a walk from Ringsend, along the strand of the Sandymount coast, to the salt marshes of Booterstown, where the Dublin Mountains loom on the horizon. It begins with the speaker crossing the hunch-backed stone bridge that spans the Dodder River:

> Whereas dart away through the cavorting scapes
> bucket o'er Victoria Bridge that's the idea
> slow down slink down the Ringsend Road
> Irishtown Sandymount puzzle find the Hell fire.[45]

Thanks to a GIStimeline (figure 7.8), we can visualize the places that anchor Beckett's early prose and poetry along with his shifting, idiosyncratic perceptions moving through the various streetscapes and landscapes of Dublin and its hinterlands.

London, 1933–35

Centered on the Irish emigrant experience of London in the early 1930s, *Murphy* (1938) culminates in the denouement of Beckett's protagonist after he takes up employment in a lunatic asylum to prevent his fiancée, an erstwhile prostitute named Celia, from resuming her profession. A subplot revolves around a cast of Irish characters traveling from Dublin to London in vain pursuit of Murphy who dies because of a gas explosion in his garret at the Magdalen Mental Mercyseat before they can locate him. In *Murphy,* Beckett is preoccupied with lampooning the rational Cartesian framing of space and experience, along with its construction of identity as a self-contained thinking subject (or *Cogito*), which posits a mind-body split. Accordingly, a distinct contrast exists between the geographical settings of the novel, which are firmly situated in Cartesian space, and Beckett's representation of Murphy's perception of the outer world that constitutes time and place as a "big blooming buzzing confusion" as he moves through the streets of London and the wards of a lunatic asylum.[46] Cartesian metaphors in Beckett's narrative consist of garrets, pens, prisons, asylums, and padded cells. Beckett foreshadows the contrasts between rational and social perspectives of space early in the novel in a set piece in which Murphy's guru Neary, a proprietor of the Pythagorean Academy of Meditation in Cork admonishes his pupil: "Murphy, all life is figure and ground" and receives an enigmatic reply: "But a wandering to find home."[47] Murphy then flees to London, which Beckett charted after consulting the 1935 edition of *Whittaker's Almanac* and by long walks taken from his lodgings in Paulson Square and Gertrude Street through Chelsea and West Brompton to the Thames Embankment. Murphy's wanderings occur not only within his mind but also across a geography of London that is plotted with a cartographer's precision.

Impressions Beckett gathered during his walks ground the embodied performances of the human landscape spanning the Battersea and Albert bridges (figure 7.9), where Celia takes a respite from her trade:

> She walked to a point about halfway between the Battersea and Albert Bridges and sat down in a bench between a Chelsea pensioner and an Eldorado hokey-pokey man, who had dismounted from his cruel machine and was enjoying a short interlude in paradise. Artists of every kind, writers, underwriters, devils, ghosts, columnists, musicians, lyricists, organists, painters and decorators, sculptors and statuaries, critics and reviewers, major and minor, drunk and sober, laughing and crying, in schools and singly, passed up and down. A flotilla of barges, heaped high with waste paper of many colours, riding at anchor or aground on the mud, waved to her from across the water. A funnel vailed to Battersea Bridge.[48]

In terms of language, "a set of proper nouns signifying manufacture, labor, the force of law, slaughter, and commerce" compose Beckett's portraits of streetscapes.[49] Spaces of capital and industry, such as the Vis Vitae Bread Co. and The Marx Cork Bath Mat Manufactory, as well as the location of the apartment that Celia and Murphy share on Brewery Road between Pentonville Prison and the Metropolitan Cattle Market, "anchor a hard, and alienating reality."[50]

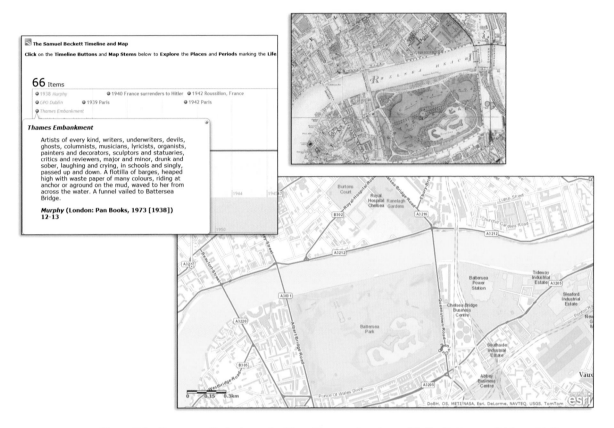

Figure 7.9 **Battersea Park along the River Thames, London, with the Battersea Bridge at left, Chelsea Bridge at right, and Albert Bridge between them. A corresponding historical map and a GIStimeline are also shown.** Courtesy of Trinity College Dublin. Data from ArcGIS Online, courtesy of Esri, HERE, DeLorme, TomTom, Intermap, increment P Corp., GEBCO, USGS, FAO, NPS, NRCAN, GeoBase, IGN, Kadaster NL, Ordnance Survey, Esri Japan, METI, Esri China (Hong Kong), swisstopo, MapmyIndia, © OpenStreetMap contributors, and the GIS User Community. Historical map of Battersea Park from Stanford's Library Map of London and its Suburbs, 1891, by Edward Stanford.

France, 1945–46

The eruption of the Second World War in 1939 caused the Irish Free State to adopt a policy of neutrality, designated officially as the "Emergency." In response, Beckett returned to Paris and joined the French Resistance in response to the Nazi treatment of Jewish friends and colleagues. An informer caused him to take refuge from the Gestapo in the village of Roussillon in Vichy France until 1945. At the end of the war, Beckett volunteered with the Hospital of the Irish Red Cross at Saint-Lô in Normandy. He returned to Paris in January 1946 and, embarking on a frenzy of writing, switched from English to French as a means to strip language to the essential of his vision.[51] He later explained, "*Parce qu'en francais c'est plus facile d'écrire sans style* (In French it's easier to write without style)."[52] It has been noted that Beckett's first post-war fictions take place in the dissolution of the city as the ordering, social matrix of narrative mimesis and are structured by them.[53] His characters are marked by a paratactic, associative wandering through spaces and a social world left behind, in ruins, rejected, forgotten, unbelievable, and canceled out.[54]

Beckett's short story *Le Fin* (The End) transposes the cityscapes of Saint-Lô and Paris on Dublin and provides a map of a post-war landscape simultaneously alien and familiar. He commenced writing the story in English but completed it in French. The desolate beauty of the poem *Saint-Lô* and the dislocation that characterizes Beckett's story *Le Fin* intimates the firebombing of the town on July 25, 1944, during the Allied invasion of Normandy.

Occupied by the German Army at the time, Saint-Lô sits on the Vire River and once served as a landmark for the high-altitude Allied bombing raids of *Operation Cobra*. The operation's bombers pounded their target area with elemental fury and saturated the town with 50,000 general-purpose and fragmentation bombs.[55] In 1946, Beckett prepared a radio broadcast, *The Capital of Ruins,* for Radio Telefis Eireann documenting its post-war reconstruction. He reported that "Saint-Lô was bombed out of existence in one night. German prisoners of war, and casual labourers attracted by the relative food plenty, but soon discouraged by housing conditions,

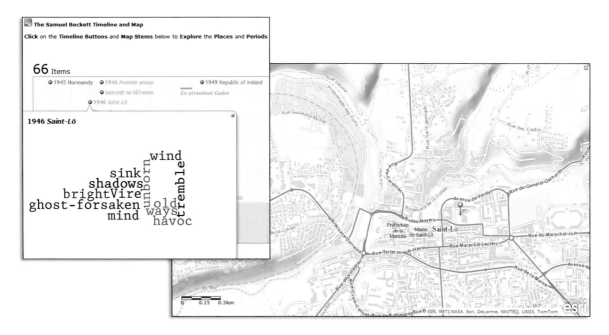

Figure 7.10 Map of Saint-Lô, France, and corresponding GIStimeline. Courtesy of Trinity College Dublin. Data from ArcGIS Online, courtesy of Esri, HERE, DeLorme, TomTom, Intermap, increment P Corp., GEBCO, USGS, FAO, NPS, NRCAN, GeoBase, IGN, Kadaster NL, Ordnance Survey, Esri Japan, METI, Esri China (Hong Kong), swisstopo, MapmyIndia, © OpenStreetMap contributors, and the GIS User Community.

continued, two years after the liberation, to clear away debris, literally by hand."⁵⁶ Beckett distilled the remains of Saint-Lô into a short poem (figure 7.10):

> Vire will wind in other shadows
> unborn through the bright ways tremble
> and the old mind ghost-forsaken
> sink into its havoc.⁵⁷

The lines echo the theme of his radio address by tracing the path of the River Vire winding through the town's apocalyptic landscape. As such, a phenomenological counterpoint to the quotidian objectivity of Beckett's piece of broadcast journalism can be intuited in the poem's:

> tiny structure: two lines about Saint-Lô, two lines about the speaker, the halves of the poem, by a baffling geometry, at once parallel and divergent. Cities, its theme runs, are renewed like rivers; men die. The Vire is a Heraclitean stream with a future of self-renewal; and the bombed city too will be rebuilt and cast shadows again. The mind [. . .] unlike the river will grow old, and will 'sink' and its 'havoc' unlike the city's will precede no second rising.⁵⁸

Beckett narrates *Le Fin* from the perspective of an existential figure with "[a] mask of dirty old hairy leather, with two holes and a slit," expelled, rather than released, from a charitable institution.⁵⁹ He finds on his release that "the city had suffered many changes, nor was the country as I remembered it."⁶⁰ Authorial perspectives of Cartesian verisimilitude that characterized Beckett's earlier depictions of cities gave way impressions of existential and phenomenological dislocation:

> In the street I was lost. I had not set foot in this part of the city for a long time and it seemed greatly changed. Whole buildings had disappeared, the palings had changed position and on all sides I saw in great letters, the names of tradesmen I had never seen before and would have been at a loss to pronounce. There were streets where I remembered none, some I did remember had vanished and others had completely changed their names. The general impression was the same as before.⁶¹

The city, built on the mouth of a bay, with two canals and mountains to the south, suggests Saint-Lô, Paris, London, and Dublin simultaneously in Beckett's prose: "the general appearance of the river flowing between its quays and under its bridges, had not changed. Yes the river still gave the impression it was flowing in the wrong direction."⁶² The vestigial landscape of 1930s Ireland, marked by the social and economic blight brought on by the Free State's economic war with Britain, appears and fades in Beckett's exposition on the human and physical desolation he encountered along the banks of the Vire among the bombed-out ruins of Saint-Lô.

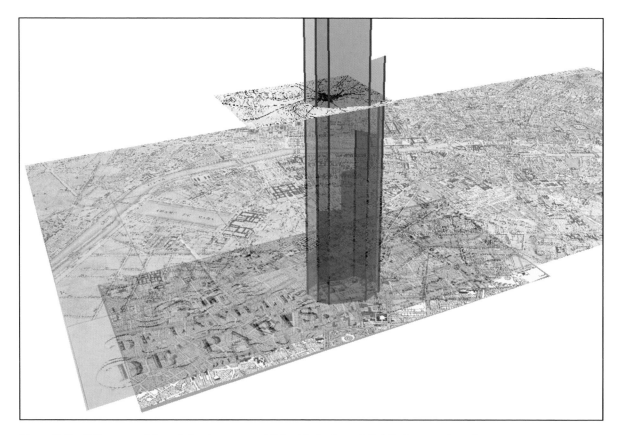

Figure 7.11 **GIS transposition of the cityscapes of Saint-Lô, Paris, and Dublin (centered on the space of Trinity College Dublin).** Imagery courtesy of Trinity College Dublin; Saorstát Eireann Ordnance Survey (OS) Dublin & Environs 1: 20000 Sh 265b, published in 1934. Obtained from Trinity College Dublin Library. De La Ville De Paris Map out of copyright; The Battle of St Lo, 11-18 July 1944 Map, source: http://ftp.ibiblio.org/hyperwar/USA/USA-E-Breakout/index.html.

The ArcScene visualization in figure 7.11 transposes three digitized historical maps over the space of Trinity College Dublin to suggest the layers of spatial memory that may have accumulated in Beckett's imagination as he was composing his story.

For Beckett, the setting of the post-war city does not function as a representation of social chaos; instead, the greatly changed city acts as a space of narrative debility.[63] This alteration is enhanced by his protagonist's incomprehension of his surroundings: "the eyes rose to a confusion of low houses, wasteland, hoardings, chimneys, steeples, and towers."[64] The story concludes with an implosion of his personal narrative perspective: "The sea, the sky, the mountains and the islands closed in and crushed me in a mighty systole, then scattered to the uttermost confines of space."[65] Beckett distilled his experience of Saint-Lô and Paris, coupled with his memory of

Figure 7.12   **1916 Easter Rising, Dublin. The flag on the map locates the center of the rebellion at the General Post Office, and the pin identifies the vantage from where Beckett and his father could see the city in flames. The accompanying GIStimeline, image, and text help readers place Beckett's childhood experience.** Imagery courtesy of Trinity College Dublin. Data from ArcGIS Online, courtesy of Esri, DigitalGlobe, GeoEye, i-cubed, USDA, USGS, AEX, Getmapping, Aerogrid, IGN, IGP, swisstopo, and the GIS User Community.

watching Dublin burn in 1916 as a child, in the story's last image of impending destruction (figure 7.12)

> It was evening, I was with my father on a height, he held my hand. I would have liked him to draw me close with a gesture of protective love, but his mind was on other things [. . .] And on the slopes of the mountains, now rearing its unbroken bulk behind town, the fires turned from gold to red, from red to gold.[66]

During the failed 1916 Easter Rising (or Easter Rebellion), Irish revolutionaries occupied the General Post Office as battles against British forces left parts of Dublin in ruins. The subsequent execution of the uprising's leaders turned public opinion toward supporting the Irish War of Independence (1919–21).

One scholar observed, "The idea that time goes through repetitious patterns is implicit in Proust, but in Beckett it is quite elaborately developed."[67] *Le Fin*, marked by Beckett's return to the latent memory of Dublin burning in 1916, effects a dissolution of Cartesian perspective and a transition in narrative from third-person English to first-person French, leaving readers with the indelible image of the storyteller whose primitive, suffering existence has become itself an emblem of the ruins into which conventional stories had crumbled.[68]

Furthermore, a GIStimeline-enabled "mash-up" that ergodically and deformatively visualizes and animates Beckett's work collates the fragmented pieces from his early prose and poetry. It charts the drift in his perspective on place from Cartesian verisimilitude to the phenomenological impressions that characterize his later, and perhaps better known, work.

Bricolage and biography

In *Proust,* Beckett characterized time as "the perpetual exfoliation of personality."[69] Both *saw* human reality as fragmentary.[70] An open source–enabled GIStimeline can make it possible to illustrate that:

> it is montage that converts space into place, that allows for place-making to occur. Place is made through narration of and for the subject, not a whole coherent image, but a sutured, fragmented image unified into a cohesive form. Place-making therefore is constructed out of the gaze, the look, and point-of-view shots.[71]

Peering through the interactive GIStimeline lens, it can be seen that Beckett's early works exhibit a latent Cartesian verisimilitude. The urban perspective in *More Pricks Than Kicks* depicts the moods and feelings of Dublin's streetscapes, anchored by a modernist sensibility that perceives the bourgeois metropolis as an organizing principle and space for fictional narratives. The Cartesian comedy of interwar London in *Murphy* depicts the protagonist's attempts to retreat from the "malignant proliferations of urban tissue," the "miasma of laws," and the "mercantille gehenna" that make the city of London.[72] In contrast, the post-war fragility of the poem *Saint-Lô* heralds the emergence of the existential Beckettian landscape, a narrative setting in which identity and place have been dislocated from the modernist metropole and located in the fragmented geographies of alienated streetscapes, ditches, rooms, and ruined cities. Indices of repetitious time and dislocated space in Beckett's post-war landscape emerge as a post-structuralist cartographer's accretion and transposition of Dublin, London, Paris, and Saint-Lô in Normandy. Splicing Beckett's literary impressions into a bricolage of biographical information using GIS techniques illuminates the idea that:

> Visualization is thus a cognitive process of learning through the active engagement with graphical signs that make up the display, and it differs from passive observation of a static scene, in that its purpose is also to discover unknowns, rather than to see what is already known. Effective geographic visualization should reveal novel insights that are not apparent with other methods of presentation.[73]

Using a GIStimeline to write biospatial narratives allows users to visualize montages of literary and biographical space side by side. To reiterate, as Beckett himself informs us, "an impression is for the writer what an experiment is for the scientist—with this difference, that in the case of the scientist the action of the intelligence precedes the event and in the case of the writer follows it."[74]

Sources

1 D. Weisberg, *Chronicles of Disorder: Samuel Beckett and the Cultural Politics of the Modern Novel* (Albany: State University of New York Press, 2000), 43.

2 See Samuel Beckett page, in *Digital Literary Atlas of Ireland*, 1922–49, http://www.tcd.ie /trinitylongroomhub/digital-atlas/writers/samuel-beckett.

3 C. Lukinbeal, "Mobilizing the Cartographic Paradox: Tracing the Aspect of Cartography and Prospect of Cinema," in *Educação Temática Digital, Campinas,* 11, no. 2 (2010): 23.

4 Dodge et al., "The Power of Geographical Visualizations," 3.

5 A. Thacker, *Moving through Modernity: Space and Geography in Modernism* (Manchester: Manchester University Press, 2003), 5; A. Thacker, "The Idea of a Critical Literary Geography," *new formations,* 57 (Winter 2005–6): 63.

6 Staley, "Finding Narratives of Time and Space," 45.

7 Thacker, "The Idea of a Critical Literary Geography," 63.

8 Ibid.

9 S. Aitken and J. Craine, "Guest Editorial: Affective Geovisualizations," *Directions Magazine* (2006), http://www.directionsmag.com/articles/guest-editorial-affective-geovisualizations/123211.

10 Staley, "Finding Narratives of Time and Space," 45.

11 Dodge et al., "The Power of Geographical Visualizations," 9.

12 S. Beckett, *Proust and Three Dialogues with Georges Duthuit* (1931; repr. London: Calder and Boyars, 1970), 84.

13 Aitken and Craine, "Affective Geographies and GIScience," 144.

14 R. Cochran, *Samuel Beckett: A Study of the Short Fiction* (New York: Twayne Publishers, 1991), 92.

15 H. Kenner, *A Reader's Guide to Samuel Beckett* (New York: Farrar, Straus and Giroux, 2003), 62.

16 Staley, "Finding Narratives of Time and Space," 45.

17 J. Pilling, *Samuel Beckett* (London and New York: Routledge & Kegan Paul, 1976), 1.

18 N. Thrift, "Intensities of Feeling: Towards a Spatial Politics of Affect," *Geografiska Annaler,* series B, *Human Geography,* 86, no. 1 (2006): 57.

19 Aitken and Craine, "Guest Editorial: Affective Geovisualizations."

20 S. Beckett, *Disjecta: Miscellaneous Writings and a Dramatic Fragment*, ed. Ruby Cohn (New York: Grove Press, 1984), 74.

21 L. Falk, "The Ethics of Perception," *Convergence,* 6, no. 1 (March 2000): 29–38.

22 D. C. Pocock, *Humanistic Geography and Literature: Essays on the Experience of Place* (Lanham, MD: Barnes & Noble Imports, 1981), 15.

23 Ibid.

24 S. A. Nathan, *Samuel Beckett* (New York: Hillary House Publishers, 1969), 28–29.

25 J. Knowlson, *Damned to Fame: The Life of Samuel Beckett* (London: Bloomsbury, 1996) 96; E. Webb, *Samuel Beckett: A Study of His Novels* (Seattle and London: University of Washington Press, 1970), 25.

26 J. J. Mayoux, *Writers and Their Work: Samuel Beckett,* no. 234, ed. I. Scott-Kilvert, (Harlow: British Council/Longman Group, 1974), 4–5.

27 L. Gordon, *The World of Samuel Beckett: 1906–1946* (New Haven and London: Yale University Press, 1996), 34.

28 Ibid., 32.

29 Ibid.

30 Knowlson, *Damned to Fame*, 126.

31 Ibid, 137.

32 D. Bair, *Samuel Beckett: A Biography* (London: Vintage, 1990), 169.

33 Beckett, *Proust*, 12, 40–41.

34 Beckett, *Disjecta*, 26–28.

35 Bair, *Samuel Beckett: A Biography*, 172.

36 Samuel Beckett, *Dream of Fair to Middling Women* (Dublin: The Black Cat Press, 1992), 101.

37 Ruby Cohn, *Back to Beckett* (Princeton: Princeton University Press, 1973), 24.

38 M. Robinson, *The Long Sonata of the Dead: A Study of Samuel Beckett* (New York: Grove Press, 1996), 75.

39 B. Singer, "Modernity, Hyperstimulus, and the Rise of Popular Sensationalism," in *Cinema and the Invention of Modern Life*, eds. L. Charney and V. R. Schwartz (Berkeley: University of California Press, 1995), 72–73.

40 S. Beckett, *Poems 1930–1939* (London: Calder, 2002), 15–16.

41 A. Proulx, "Dangerous Ground: Landscape in American Fiction," unpublished paper (courtesy of author, 2004), 8.

42 S. Beckett, *More Pricks Than Kicks* (1934; repr. New York: Grove Press, 1972), 48–49.

43 Ibid., 47.

44 Ibid., 49.

45 Beckett, *Poems*, 27.

46 S. Beckett, *Murphy* (1938; repr. London: Pan Books, 1973), 6.

47 Ibid.

48 Ibid., 12–13.

49 Weisberg, *Chronicles of Disorder*, 36.

50 Beckett *Murphy*, 40; Weisberg, *Chronicles of Disorder*, 36.

51 Knowlson, *Damned to Fame*, 358; Cohn, *Back to Beckett*, 58–59.

52 Ibid.

53 Weisberg, *Chronicles of Disorder*, 64.

54 Ibid., 68.

55 J. Sullivan, "The Botched Air Support of Operation COBRA," *Parameters* (March 1988): 98–99.

56 S. Beckett, "Capital of Ruins," in *The Beckett Country: Samuel Beckett's Ireland*, ed. E. O'Brien (Dublin: The Black Cat Press, 1986), 337.

57 Beckett, *Poems*, 34.

58 Kenner, *A Reader's Guide to Samuel Beckett,* 46.

59 S. Beckett, "The End," in *First Love and Other Novellas* (London: Penguin, 2000), 22.

60 Ibid., 18.

61 Ibid., 12.

62 Ibid., 13.

63 Weisberg, *Chronicles of Disorder*, 67.

64 Beckett, *The End*, 27.

65 Ibid., 31.

66 Ibid., 30.

67 Webb, *Samuel Beckett: A Study of His Novels*, 31.

68 Weisberg, *Chronicles of Disorder*, 65.

69 Beckett, *Proust*, 13.

70 Webb, *Samuel Beckett: A Study of His Novels*, 30.

71 Lukinbeal, "Mobilizing the Cartographic Paradox," 19.

72 Beckett, *Murphy*, 47.

73 Dodge et al., "The Power of Geographical Visualizations," 3.

74 Beckett, *Proust*, 84.

Part 3

Toward a humanities GIS

Chapter 8

The *terrae incognitae* of humanities GIS

The lost mapmaker

Parsing the word *technology* by distinguishing its Greek roots—τέχνη (téchnē), meaning "art, skill, and craft," and λογία (logía), "the study of"—illuminates new paths for GIS scholarship. We can now chart, navigate, survey, and, in turn, represent the *terrae incognitae* emerging in the wakes left by spatial and cultural turns in arts and humanities scholarship over the past century. Reflecting on my dual interests in the arts and humanities and experimental GIS methodologies and techniques, I wonder if I am similar to José Arcadio Buendía, the nineteenth-century Colombian patriarch of Gabriel García Márquez's novel, *One Hundred Years of Solitude*. Aided with Portuguese maps, an astrolabe, a compass, and a sextant, Buendía calculates in the nineteenth century what Eratosthenes first discovered in 240 BC—that "the earth is round, like an orange."[1]

Trends in human and cultural geography over the last half century have acted as midwives to the birth of geohumanities and the spatial humanities. Bertrand Westphal's notion of geocriticism, introduced in chapter 1, is one of the prominent fruits of this interdisciplinary cross-pollination. Organizing his work around geographic sites, rather than solely on authorial perspectives, Westphal focuses on the heterogeneous, multifocal views of place generated by both insider and outsiders. His work engages nonliterary texts, such as tourist brochures, official records, and reports, to explore the full spectrum of particular geographical sites. Geocriticism aims to situate real places and fictional spaces as postmodern lenses that focus on a wide range of cultural, theoretical, and aesthetic forms. Collating texts from different historical periods, Westphal employs a "stratigraphic" reading of place by mobilizing distinct but compatible methodologies.[2] The obvious potential of GIS methodology for geocritical studies could also apply to Franco Moretti's and Barbara Piatti's work on literature and cartography. The preceding chapters illustrate how historical maps may be employed in such a manner. But perhaps to facilitate these type of arts and humanities GIS engagements, we need to consider Edward S. Casey's argument that the art of mapping "needs to be liberated from its alliance with modern cartography so that it can resume its original sense of charting one's way into a given place or region."[3]

The map theater

Michel de Certeau's "Map Theater" provides a metaphor to reflect on the possibilities for reconceptualizing and reconfiguring GIS methodologies for the humanities, as an interpretive tool rather than an embodiment of complete knowledge or as a finished product. As we peer into the digital "proscenium" of the computer screen, a visual perspective established in the Renaissance focuses our gaze, like that of a theater patron. GIS stories therefore constitute a codescape of computer language and commands. GIS visualizes a geographical syntax and scripts its operations, à la Shakespeare, like lines in a Cartesian wordplay. William Cartwright observes that by "using the concept of the Storyteller, and applying it through the metaphor of the Theatre cinema can be used to enhance 'map' information and provide a richer story of a landscape or a culture."[4]

The digital revolution of the twenty-first century profoundly reconfigured geospatial studies by bringing this "Map Theater" under the control of our fingertips. Scholars in the arts and the humanities are currently scripting and developing cutting-edge techniques and innovations using GIS and other cybernetic platforms. However, as Willard McCarthy recognizes, computational models produced by such efforts, "however finely perfected, are better understood as temporary states in a process of coming to know rather than fixed structures of knowledge."[5] McCarty reminds us that "for the moment and the foreseeable future, then, *computers are essentially modeling machines, not knowledge jukeboxes.*"[6]

Simply stated, GIS is not "geography in a box." Like the Trojan horse set before the gates of Troy, geospatial software is now being invited into the houses of the arts and humanities. There it sits, heir to an ancient and established academic practice. However, danger abounds! Soldiers of a Euclidian epistemology are infiltrating the historical, literary, cultural, and artistic courtyards of knowledge and destroying the aesthetic and hermeneutical palimpsests of time and place so tenderly cultivated by scholars in these disciplines.

Attempting to survive the throes of the digital revolution now reconfiguring the academic landscape, scholars generally (with exceptions, of course) seem to engage geospatial technology in one of two ways. Either they establish a division of labor between "humanists" and "technicians," or they manipulate software programs to perform quantification exercises and flout algorithms. The trouble with the former is that GIS is perceived as a means to produce a finished product, rather than as a tool of iterative analysis. The scholar remains geospatially unaware and dependent on the technician. The danger with the latter is that scholars may find themselves in a cul-de-sac of informatics, without grasping how geographical perspective and critical spatial thought, coupled with emerging geovisualization and GIS techniques, can multi-dimensionally illuminate their field of study and provide a sophisticated means with which to conduct spatio-discursive analysis.

Denis Wood, John Fels, and John Krygier contend that using an open-source geospatial platform such as "Google Earth may feel like magic, but it's not, or it's the magic of a Fred Astaire dance, effortless only because so long rehearsed."[7] Without some awareness of the geographical tradition, proprietary GIS software fosters a similar, albeit more sinister and panoptic illusion. Gunnar Olsson warns,

> The tools order us what to do. It is not we who rule the tools, but you end up often with
> the tools ruling us . . . And then there is a temptation to overuse the tools, to (over) extend

them. So you have to know very well when to stop. The trick is not to stop, but to know when to stop.[8]

Elder jazz musicians tutor their pupils in the same manner; as Count Basie said, it's the notes you *don't* play that count. GIS critiques, Eric Sheppard notes, focus on the danger that the technology will overpower post-positivist perspectives and promote purely empiricist epistemologies. Conventional GIS approaches inadequately represent either non-Western conceptions of space or the communicative rationality and irrationalities of everyday life.[9] Furthermore, GIS applications reinforce certain conceptions of geometric and relative space and types of reasoning (Boolean logic in particular). Giles Deleuze drew up a list of spaces— geometrical, physical, biophysical, social, and linguistic—that GIS specialists can consider when formulating methodologies that address these critiques.[10]

David Bodenhamer argues that the technology has the potential to further revolutionize the role of place in the humanities. Innovators are creating a new experiential, rational base of knowledge by moving beyond 2D map images in order to represent and explore dynamic and interactive systems.[11] Whether GIS becomes viable and sustainable in these disciplines and harnesses the ontological and epistemological strengths of the arts and humanities, however, depends on the willingness of its users to reboot. Here, the roles of metonymy, subjectivity, and perception demand consideration. Analysts should take critical and literary theory into account when studying GIS epistemologies—how narratives are constructed and incorporate tropes from art and film history, for example. Perhaps considering GIS as a digital-canvas and textual-analysis platform, a cinematic cartographic editing suite, and a postmodern type of video game that scholars can hack to conduct textual analysis will aid in reconfiguring its use. With its Python scripting capability, GIS can build a poetic bridge between the realms of literature, linguistics, art, and computer science. Other such disciplinary linkages just have to be imagined. Observe the spectrum of complex world systems that range in scale from a protein molecule unfolding in the nuclei of a cell to the roiling, nonlinear dynamics of the earth's atmosphere, for example. Pondering this, David Porush argues,

> Reality exists at a level of human experience that literary tools are best, and historically most practiced, at describing . . . [therefore] by science's own terms, literary discourse must be understood as a superior form of describing what we know.[12]

The GIS images, maps, and visualizations featured in this book were created with commensurate tools in mind. They do not adhere to the exact rules of classical cartography. Nor would their color choices, use of symbols, and presentations likely pass muster in a conventional undergraduate map-design class. Although GIS serves fundamentally as a cartographic tool, the potential of the technology extends far beyond the cartographical pale.

And that is the underlying point of this book.

Mapping out unknown lands, attempting to visualize critical theory, and reimagining (or *Imagineering,* as the folks at Disney like to put it) how "three key referencing systems—space, time, and language—might be engineered in such a way that changes in one ripple into the others"[13] involve some creative "crayoning" outside the lines that demarcate standard cartographic practices and transgression across established GIS epistemological lines.

The geographer's science and the storyteller's art

Stuart Aitken and James Craine proclaim that, with the advent of digital domains, geographic knowledge "has given a new, and ontologically different, life to geography."[14] They argue that "we now have a new realm of images with extended, but still valid, reconstructions of the 'real.'"[15] In turn, perspectives from the arts and humanities can illuminate new paths to pursue using GIS. As Gunnar Olsson points out,

> For what is that type of mapping at a distance if not a human activity located in the interface between poetry and painting? What is a satellite picture if not a peephole show, a constellation of signs waiting to be transformed from meaningless indices into meaningful symbols?[16]

GIS provides access to undiscovered sites for geographic discovery.[17] Such sites fill the arts and humanities landscape and offer new insights for further exploration to cognate disciplines. For example, discussing how art can inform the scientific practices of mapping, Olsson looks to the painter Cézanne and the poet Mallarmé, who were "aware that if it were not for the infinite plasticity of paint and words they could never have invaded the silences that normally fall outside the limits of grammar."[18] Olsson acknowledges that modern techniques of remote-sensing and aerial photography could benefit from the technical skills and theoretical insights of these two artistic maestros.[19] What other artists, poets, and writers in the canons of Western history (and perhaps non-Western history) can educate our arts and humanities GIS practices? Perhaps, as Olsson suggests, we can grasp through reverse "imagineering" how GIS stands naked before us, shamefully parading as a game of ontological transformations in which theory-laden observations are translated first into patches of color, then into strings of words, and finally into purposeful action. Picture becomes story, and "is" turns to "ought."[20]

I believe that GIS scholarship in the arts and humanities is spawning its own unique language, tools, perspectives, methodologies, and storytelling techniques as it adheres to the dictates and tropes of its cognate disciplines. Borrowing from Deleuze, we can reimagine GIS software for the arts and humanities as the painting machine of an artist-mechanic, which clears the canvas (or the computer screen, web page, or data cloud) of all the clichés that prevent innovative creation.[21] Just as David Bowie and Brian Eno adopted and adapted synthesizers, we should interact with GIS but not necessarily in the way it has been conventionally designed and applied. Instead, we should explore the technology to ascertain what sort of toolkits it offers arts and humanities researchers. There is ample evidence that such considerations are being entertained, and the development of such approaches is underway. For example, the digital historian Alexander von Lünen, echoing Deleuze, argues,

> Rather than a visualization tool, GIS should be used as a painting tool; a tool to creatively engage with one's sources. Historians need to make a decision: will they remain in the "painting by numbers" domain or will they develop into true GIS artists, abandoning the imitation game and transforming GIS into a genuine vehicle for historic inquiry?[22]

Concerning GIS approaches for the arts and humanities, the aphorism "if a hammer is your only tool, all your problems are going to look like nails," seems apropos.[23] New scholarship should not be shoehorned to fit how GIS has been engineered; instead, GIS manuals "hermeneutically" consulted during research in these cognate disciplines can yield insights and unexpected discoveries. The preceding chapters in this book bear the fruits of such an approach. Arts and humanities researchers are currently transforming GIS technology by integrating its suite of tools—its digital video and audio platforms, online web services, and techniques of 3D geovisualization—to create metaphorical, metonymical, virtual, and genuinely alternative representations of time and space. Consequently, we can reimagine GIS applications as a set of integrative, hermeneutic, and artistic tools to create digital platforms, such as the GIStimeline of Samuel Beckett's early life and travels. Many of the images and visualizations in this book include a "backstage view" of the GIS "Map Theater" workspace (data view) to demonstrate how GIS can be employed to storyboard and annotate the different spatio-discursive layers of texts, as in the studies of Patrick Kavanagh, James Joyce, and Flann O'Brien. GIS also provides a means to operationalize their depictions of urbanity, illustrating, in the words of William Mitchel, how:

> Cities and buildings like films might be scored by famous composers—with the soundtrack electronically edited, on the fly, as you moved around. And everything of relevance at a particular location (for example, a historic site or a crime scene) might be retrieved and arrayed to provide a comprehensive, electronic mise-en-scène.[24]

Literary scholars can use GIS as a form of exegesis to extirpate multiple forms of space and geography that, Andrew Thacker observes, "cannot, it seems be kept apart, even though there is for many writers, a desperate desire to maintain borders and boundaries: rooms bleed into streets, anguished minds migrate to lands overseas."[25]

The historical GIS case study in chapter 3 explored the 1641 Rebellion in Ireland, Cromwell's subsequent invasion, and William Petty's cartographic techniques that were used to tabulate the seventeenth-century *Books of Survey and Distribution*. The *Books*, in Audrey Kobayashi's phrase, can be considered as English "statistexts"[26] that reconfigured the political-economic Irish landscape. GIS databases parsed the *Books* to map and visualize land redistribution patterns. In this regard, the GIS interface was used as a storyboard to organize data layers in an iterative research process called *landscapification*, which "codes the world, imposing physical and psychological hierarchies."[27]

These and other arts and humanities GIS projects offer exciting pedagogical possibilities for introducing students to develop cross-disciplinary approaches of their own. Jennifer Lund and Diana Stuart Sinton state that GIS mapping lies at the intersection of the arts and the sciences. While supporting intuitive perceptions and deductive analysis, GIS liberates students to be provocative and permissive by putting information together in new ways and allowing unjustified arrangements of information. Students can use GIS to critically and creatively break up packaged units of perspectives and the premises imposed on themselves and others.[28] This point of view is analogous to the visual aims of the Impressionists, German Expressionists, Italian Futurists, Dadaists, and Surrealists. Discussing their geospatial "translation" of Charles Baudelaire's performance as a *flâneur* and his poem, *Les Hiboux* (1857), Gwilym Eades and Renée Sieber observe,

> Metonymical language is primarily the language of science, while metaphorical language is associated with art [...] while both art and science have equal claims to epistemological

validity, in GIS environments, we would suggest that the technical object of GIS is associated more with science and thus with metonymy. The introduction of art (and even "the art of cartography") into GIS . . . in the form of metaphor, provides a useful counterbalancing and corrective.[29]

Eades and Sieber conclude, "We believe that technology holds liberating potential, but technology must not be conflated solely with science. It must also ally itself with art. As Heidegger . . . pointed out it is only when we start to think of technology as neutral that we deliver ourselves over to it in the worst possible way."[30]

A recent *New York Times* article titled *Digital Keys for Unlocking the Humanities' Riches* argued that the arts and humanities have entered a post-theoretical age and methodological moment "similar to the late 19th and early 20th centuries, when scholars were preoccupied with collating and cataloging the flood of information brought about by revolutions in communication, transportation and science."[31] Major research libraries are at the forefront of employing GIS as a tool in this endeavor. But GIS is more than just a cataloging tool. Combined with theory and focused through the disciplinary lens of the arts and humanities, GIS allows us to survey, chart, navigate, and explore unknown territories that may, in the end, alter and transform our ability to perceive our worlds.

Eratosthenes, a father of Western geography, was the head librarian of the Great Library at Alexandria when he made his discovery that the earth was indeed a sphere, raising human consciousness to a state that fundamentally altered the course of history. Over the past decade, Denis Cosgrove observes, thinking in science and technology studies has begun to dissolve epistemological distinctions between the arts and the sciences. Just as science's claims to a universal truth diminished in the face of postmodernity, modern artists have been rejecting aesthetics as the defining feature of their work.[32] Bruno Latour once pointed out that it was much more difficult to extirpate science from its epistemological past than to free art history from aesthetics. However, once both disciplines were liberated from their epistemological and aesthetic constraints, a vast common ground opened up, and visualization in science and the visual arts became truly vascularized.[33]

Such epistemological and aesthetic shifts suggest that perhaps the spiral staircase of history is leading us to an ontological plateau where we will rediscover, as the ancient Greeks and Romans did before us, that the geographer's science and the storyteller's art are standing again on common ground—albeit with the aid of humanities GIS.

Sources

1   Gabriel García Márquez, *One Hundred Years of Solitude* (New York: Harper Collins, 1998 [1967]), 5.

2   B. Westphal, *Geocriticism*; R. T. Tally, Jr., *Geocritical Explorations*.

3   E. Casey, "Boundary, Place, and Event in the Spatiality of History," *Rethinking History*, 11, no. 4 (2007), 512.

4   W. Cartwright, "Applying the Theatre Metaphor," 28.

5   W. McCarty, "Modeling: A Study in Words and Meanings," in *A Companion to Digital Humanities*, ed. Susan Schreibman et al. (Oxford: Blackwell, 2004), http://www.digitalhumanities.org/companion.

6   Ibid.

7   D. Wood, with L. Fels and J. Krygier, *Rethinking the Power of Maps* (New York: Guilford Press, 2010), 17.

8   G. Olsson and A. von Lünen, "Thou Shalt Make No Graven Maps!" in *History and GIS: Epistemologies, Considerations and Reflections*, eds. A. von Lünen and C. Travis (New York: Springer, 2012), 79.

9   E. Sheppard, "Knowledge Production through Critical GIS: Genealogy and Prospects" *Cartographica, 40*, no. 4 (2005): 7.

10 G. Flaxman, "Transcendental Aesthetics: Deleuze's Philosophy of Space," in *Deleuze and Space*, eds. I Buchanan and G. Lambert (Edinburgh: Edinburgh University Press, 2005), 187.

11 D. J. Bodenhamer, "Beyond GIS," 1–14.

12 D. Porush, "Progogine, Chaos and Contemporary SF," *Science Fiction Studies,* 18, no. 3 (1991): 77.

13 Corrigan, "Qualitative GIS and Emergent Semantics," 85.

14 Aitken and Craine, "Affective Geographies and GIScience," 144.

15 Ibid.

16 G. Olsson, *Abysmal* (Chicago: University of Chicago Press, 2007), 137–38.

17 Aitken and Craine, "Affective Geographies and GIScience," 144.

18 Olsson, *Abysmal*, 137.

19 Ibid., 138.

20 G. Olsson, "Constellations," *Sistema Terra,* 8, nos. 1–3 (1999): 141.

21 G. Deleuze, "Cold and Heat," in *Photogenic Painting, Gerard Fromanger*, trans. D. Roberts (London: Black Dog Publishing, 1999), 64; S. Zepke, *Art as Abstract Machine: Ontology and Aesthetics in Deleuze and Guattari* (London and New York: Routledge, 2005), 8.

22 A. von Lünen, "Tracking in a New Territory: Re-imaging for History," in *History and GIS: Epistemologies, Considerations and Reflections*, eds. A. von Lünen and C. Travis (New York: Springer, 2012), 237.

23 I attribute this to my mentor and colleague, Professor Emeritus Richard Scott of Rowan University.

24 W. J. Mitchel, *Me++ The Cyborg Self and the Networked City* (London: MIT Press, 2003), 123–24.

25 A. Thacker, *Moving through Modernity: Space and Geography in Modernism* (Manchester: Manchester University Press, 2003), 7.

26 R. Fiedler, N. Schuurman, and J. Hyndman, "Improving Census-Based Socioeconomic GIS for Public Policy: Recent Immigrants, Spatially Concentrated Poverty and Housing Need in Vancouver," *ACME: An International E-Journal for Critical Geographies,* 4, no. 1 (2006): 149.

27 S. Zepke, *Art as Abstract Machine*, 127.

28 J. J. Lund and D. Stuart, "Critical and Creative Visual Thinking," in *Understanding Place: GIS and Mapping Across the Curriculum*, eds. D. S. Sinton and J. J. Lund (Redlands, Esri Press, 2007), 14.

29 G. Eades and R. Sieber "Mapping the *Flâneur*: A Geospatial Translation of Charles Baudelaire's 'Les Hiboux'" (paper presentation, meeting of the Canadian Association of Geographers, Ottawa, 2009), 23.

30 Ibid.

31 P. Cohen, "Digital Keys for Unlocking the Humanities' Riches," *The New York Times*, November 17, 2010 http://www.nytimes.com/2010/11/17/arts/17digital.html?pagewanted=all&_r=0), accessed 21 January 2014.

32 D. Cosgrove, "Maps, Mapping, Modernity," 51.

33 B. Latour, "How to Be Iconophilic in Art, Science and Religion?" in *Picturing Science, Producing Art*, eds. P. Galison and C. A. Jones (London and New York: Routledge, 1998), 425.

About the author

Charles Travis is a senior research fellow with the Trinity Long Room Hub at Trinity College, Dublin, where he holds a PhD in geography. He develops methodologies and applications for humanities geographical information systems (HGIS) and conducts research in the digital and environmental humanities and in human and literary geography. He contributed to and co-edited the volume *History and GIS: Epistemologies, Considerations and Reflections* (2012), with Alexander von Lünen; and published the monograph *Literary Landscapes of Ireland: Geographies of Irish Stories, 1929–1946* (2009). Additionally, his work appears in the *International Journal of GIScience*, the *International Journal of Humanities and Arts Computing, CITY,* and *Historical Geography.* He completed a postdoctoral fellowship in the digital humanities at Trinity College and has held teaching posts at Trinity College, the University of South Florida, and several other universities.

٢

Index